MORE PRAISE FOR BEYOND THE FIRMAMENT

I have read many books over the years attempting to deal with the Bible and Science. It is rare to find one that can address both the Bible and Science well and offer straightforward discussion that cuts through the complexity of the subject. Glover's faith is communicated strongly, his understanding of the Bible's demands is well-informed, and his presentation of the scientific evidence is persuasive, concise, and remarkably clear. It is just the right kind of book to put into the hands of someone who is struggling with the subject. I have been recommending this book to everyone.

JOHN H. WALTON
Department of Biblical and Theological Studies
Wheaton College and Graduate School

The universe is billions of years old, and life on our planet is unmistakably characterized by common ancestry. For Christians, the questions are not new, but lately they seem more urgent. How then do we read the great creation accounts in Genesis? What does it mean to hold a high view of biblical revelation while honestly considering the facts of natural history? Gordon Glover has created a delightfully readable yet comprehensive exploration of the relationship between the Genesis narratives and the science of our day, focusing on the context in which those narratives were given to humankind. *Beyond the Firmament* combines clarity and wit, honoring the Word of God while respecting our understanding of God's world. This is an impressive achievement for the author and a Godsend for Christians.

STEPHEN MATHESON
Associate Professor of Biology, Calvin College
http://sfmatheson.blogspot.com

Christians who care about origins issues should definitely read *Beyond the Firmament*. It is well-written and easy to understand, and it nicely summarizes what God's Word and God's world are telling us about creation. It includes the best discussion I've read for how and why we should put ourselves in the mindset of the original author and audience of Genesis, and how doing so will help us understand the real message of Genesis so we can get beyond the debates about creation, age, and evolution.

LOREN HAARSMA
Assistant Professor of Physics and Astronomy
Calvin College

Beyond the Firmament is the place to start for Evangelicals trying to reconcile the findings of modern science with a high view of Scripture. Glover's approach is accessible without being overly simplistic. Highly recommended.

As one of the few "conversational tone" books I've read that I've thoroughly enjoyed, Gordon's *Beyond the Firmament* has a special place on my shelf. The book's greatest strengths are its down-to-earth style and Glover's unique ability to communicate scientific, theological, and historical truths in a way that no author to date has done in the creation/evolution debate. As a model of intellectually honest and loving conversation, I unreservedly recommend *Beyond the Firmament* to anyone interested in discussing this hot-button topic with friends and family.

Glover's strong points are his knack for analogy and his conversational style. His humility and honesty about the limitations of science make bearable the experience of shattering the "godless atheist" facade so many evangelicals have constructed in front of mainstream science. For anyone who might be interested in looking into the issue of evolutionary creationism, *Beyond the Firmament* is the place to start.

Beyond the Firmament is easy to read, informative, and enjoyable. Every believer who struggles with Genesis and Inspiration should read this book. I recommend it to all my readers who wish to have a clearer world-view when it comes to the sciences of origins and the Bible.

BEYOND THE FIRMAMENT

Understanding Science and the Theology of Creation

BEYOND THE FIRMAMENT

Understanding Science and the Theology of Creation

By Gordon J. Glover

WATERTREE PRESS

Beyond the Firmament: Understanding Science and the Theology of Creation

Second Printing, July 2008

Published by Watertree Press LLC
PO Box 16763, Chesapeake, VA 23328
http://www.watertreepress.com

All scripture quotations, unless otherwise indicated, are taken from the HOLY BIBLE, NEW INTERNATIONAL VERSION®. NIV®. Copyright © 1973, 1978, 1984 by International Bible Society. Used by permission of Zondervan. All rights reserved.

Scripture quotations marked NASB taken from the New American Standard Bible®, Copyright © 1960, 1962, 1963, 1968, 1971, 1972, 1973, 1975, 1977, 1995 by The Lockman Foundation. Used by permission.

Scripture quotations marked KJV are taken from the Holy Bible, King James Version, Cambridge, 1769.

Cover design by Enterline Design Services, LLC

Printed in the United States of America

Library of Congress Control Number: 2007935048

ISBN-13: 978-0-9787186-1-9
ISBN-10: 0-9787186-1-5

Publisher's Cataloging-in-Publication Data
Glover, Gordon J.
   Beyond the firmament: understanding science and the theology of creation / Gordon J. Glover.
    p. cm.
    ISBN-13: 978-0-9787186-1-9
    ISBN-10: 0-9787186-1-5
    1. Bible and science. 2. Evolution (Biology). 3. Providence and government of God—Biblical teaching. 4. Apologetics. I. Title.
BS650.G56   2007
215—dc21   2007935048

Related Websites:
http://www.beyondthefirmament.com
http://www.watertreepress.com

CONTENTS

ACKNOWLEDGEMENTS

I would first like to thank my wife and family—not only for encouraging me to take on this project, but for basically giving up an entire summer of "daddy time" so that I could capture my thoughts without too many distractions, and for sacrificing the numerous evenings, weekends, and holidays needed to get from the rough draft to a completed manuscript during the months that followed.

I am obviously indebted to Jason Baker at Watertree Press—not only for taking a chance on a first-time author for their first book release, but for taking a personal interest in the manuscript and providing invaluable editorial advice. Speaking of editorial advice—I should again thank my wife Katy, and also my mother Barbara Ferreira, my parents-in-law Pete and Kelly Hanson, Pastor David Lescalleet, Pastor Mark Scott, Erik Lytikainen, and Mike Espey for graciously reviewing parts or all the early drafts of the manuscript and helping me with the "tonal" issues; and to Tim Feeney for providing valuable last minute editing comments. Special thanks are also due to Hal Perrin for his meticulous contributions to the second printing.

I cringe to think how differently this book may have turned out were it not for the many spirited discussions I've had with random individuals on blogs, in person or by email. It would be impossible for me to list everyone who contributed to this ongoing dialogue—authors, bloggers, educators, pastors, and laypersons—but I can't overstate how important all of them were to keeping me focused on the theological issues most important to conservative Christians who, like me, take the Bible seriously.

And finally, I could not have done this without the scientific and theological expertise of those dedicated individuals in their respective fields. In addition to the many scholars I have quoted and/or referenced directly, there were many who graciously offered to review the manuscript and provide feedback, reviews, or endorsements. While I take full responsibility for any errors that remain, I am grateful for the assistance of Rev. Jack Carter, Professor David W. Opderbeck, Brian D. McLaren, Rev. P. Andrew Sandlin, Dr. Christopher M. Sharp, Dr. David Campbell, Phil Jones, and Mike Graves.

PREFACE

The relationship between science and religion is a curious thing. For instance, when I read the Bible or study theology, I recognize a certain "ultimate" order to the world. I don't always understand it, but there is a meaning and purpose to the universe that transcends our finite intellects and is beyond our own selfish inventions and devices.

This unseen spiritual economy—or "Kingdom of God" as the Bible refers to it—clearly lies outside the reach of empirical science, and yet it is no less real than the keyboard under my fingertips. While it can't be quantified or observed directly, this Kingdom manifests itself here on earth as its citizens rise above their self-serving natures and do irrational things like loving their earthly enemies, turning the other cheek for the sake of peace, rejoicing in suffering for a greater purpose, bearing one another's burdens in human solidarity, or sacrificing precious resources to promote counterintuitive concepts like *mercy* and *justice* in parts of the world that the more "successful" members of our species shouldn't even care about. The cumulative effect of this otherworldly behavior moves us slowly toward the day when it will be "on earth as it is in heaven."

So it goes on Sunday, but how do we look at the world come Monday morning? There are other ways to describe reality that seem to work well—very well, actually. For all practical purposes, the physical universe operates like a finely-tuned machine according to a strict set of unbreakable rules that seem to leave no room for supernatural or otherworldly influence. Who can deny that there are definite and discernable patterns of cause and effect that absolutely and universally dominate space, time and matter? Even our own emotions can often be tied to chemical pathways and electrical signals. How can citizens of a Kingdom that is not "of this world" navigate such a cold, impersonal and morally ambivalent universe?

This tension between the *spiritual* and the *physical* has challenged philosophers for centuries, and men have proposed various resolutions over the years. During the Enlightenment, when science overtook religion as the popular framework for understanding the universe and man's place in it, many assumed that after God had created the heavens and the earth, He simply "stepped back" and allowed the forces of Nature to take over. Every once in a while, God may still do something "miraculous" just so we thickheaded creatures don't forget that He is out there. But in general, a post-Enlightenment (modern) worldview demands that His involvement with the cosmos be very limited.

The current (postmodern) version keeps God in His box by drawing sharp lines of distinction between the spiritual and the physical realms. We can simply avoid conflict by keeping these two domains far apart from one another. When you approach it like this, the Bible is restricted to telling us only about things necessary for knowing God and the way of salvation, and science is our guide to material cause and effect. To quote the infamous Italian astronomer Galileo Galilei (1564–1642), "The Bible tells us how to go to heaven, not how the heavens go." Since the time of Galileo, many scientists—both secular and religious—have extended this olive branch to people of faith.[1]

On the surface, separating science and faith can seem like a reasonable compromise. For example, if you want to know how babies are made, you should learn how a sperm cell fertilizes an egg cell, and how their genetic material combines to make a unique blueprint for a new human life. The Bible simply has nothing to say about the mechanics of it. But if you want to know when it's permissible to hop in the sack with the opposite sex and conduct some personal research on the miracle of life, then you obviously consult the Scriptures, or (*insert religious tradition here*). Nature can't instruct us on the proper *meaning* and *sacredness* of sexual union because *meaning* and *sacredness* are transcendent[2] concepts entirely unrelated to the technical details of sexual reproduction.

That all seems fair enough, but this same line of reasoning is often used by the secular world to dismiss any claims the Bible does make about the material realm. It is not uncommon to read or hear statements like,

> ...if you believe that an adequately loving God must show his hand by peppering nature with palpable miracles, or that such a God could only allow evolution to work in a manner contrary to facts of the fossil record (as a story of slow and steady linear progress toward *Homo sapiens*, for example), then a particular, partisan (and minority) view of religion has transgressed into the magisterium of science by dictating conclusions that must remain open to empirical test and potential rejection.[3]

---

[1] For a recent example of this principle, see Stephen Jay Gould, *Rocks of Ages: Science and Religion in the Fullness of Life* (New York, NY; Ballantine, 1999).

[2] *Transcendent* is a word that I'll frequently use to refer to those things that "transcend" the material world. They are metaphysical concepts that exist independently of time, space and matter.

[3] Ibid, pg. 94.

In other words, you can have your *religion*—as long as it doesn't make any objective claims that compromise the known laws of nature. The unfortunate result is that anything miraculous is immediately struck down by the courts of science as physically impossible. The burden of scientific proof is placed on Christians to provide physical evidence for Christ's resurrection and ascension into heaven (and the hundreds of eyewitnesses don't seem to count as valid evidence). Consequently, the Bible is eventually rendered irrelevant. After all, why trust a book that contains so many obvious "misconceptions" about reality—so many outrageous and unsupported claims?

If resolving conflicts between science and the Bible were as simple as erecting a wall of separation between the physical and the metaphysical, then there probably wouldn't be much controversy in the first place. But Christianity is more than just a nice moral theory. Our faith rests on certain material facts, like the incarnation, crucifixion and resurrection of Jesus Christ, who was God in the flesh. To paraphrase the Apostle Paul writing to the Corinthian Church (1 Corinthians 15:13–19): if these historical events did not physically take place, then we are wasting our time.

Many scientists seem willing to overlook a few breaches of naturalistic etiquette strictly for the sake of peace with religion, but our faith goes far beyond just acknowledging these miracles. It provides a comprehensive biblical *worldview* that extends to every area of life—even the natural sciences. As a result, most of the controversy between science and religion stems from those passages of Scripture that appear to be making objective scientific claims about the physical universe—particularly about its formational history (how and when things came into existence). And when our biblical interpretations fail to line up with the scientific conclusions, the *creation/evolution* controversy continues. And make no mistake about it; this debate has become the front line of an epic battle over biblical inerrancy.

How did we get to this point? Most people are familiar with the story. During the mid 1800s, Darwin's theory of evolution began to replace the Garden of Eden as the popular view of man's origins within the scientific community. According to "science," no longer were supernatural acts of special creation[4] needed to account for the diversity of life on the planet. By the mid to late 1900s, the Big Bang cosmology had completely removed God from having any role in the origin of the cosmos and the emergence of celestial systems. Some scientists were more sober in their assertions about the supernat-

---

[4] The term *special creation* specifically refers to a transcendent act of creation ex nihilo (out of nothing). The term *creation* (without the term special) can refer to something created ex nihilo, or to something that was "created" naturally, by the laws of nature acting on pre-existing matter (like you and me for instance).

ural and were careful not to use these new theories as weapons against the church. Others saw evolution and the Big Bang as the last two nails in the supernatural coffin and were eager to hammer it shut forever.

Conservative Christians naturally reacted to the latter group, and rightly so. But instead of challenging the underlying assumptions by asking why any discovery of material cause and effect necessarily precludes God's creative action, most Christians challenged the claims of science on the material level, insisting that only the Scriptures can provide us accurate technical details of creation. Using the Word of God as a "pocket reference" of natural history, Christian apologists (with or without any formal scientific education) challenged the evolutionists' arguments point by point; even bypassing the established scientific peer review process to try to directly sway public opinion.

Their intentions may have been good, but all that these flagrant violations of scientific protocol accomplished was to fire up the opposition. Over time, even the moderate scientists were encouraged by this disregard of the established process to take sides against the Christian faith. Whether these moderates ever really cared about creation or evolution became secondary. Once the ethical standards of their profession were under assault, they had no choice but to join the fight to defeat the "religious zealots" who would have society return to a time when the earth was flat and Apollo's chariots pulled the sun across the firmament each day. Despite a long history of scientific progress in monotheistic cultures, religion became equated with "antiscience" and was labeled as a dangerous threat to human progress.

The religious faithful intensified their assault on evolution and the Big Bang by developing alternate scientific theories, based on the Scriptures, that demonstrated a young earth and a biosphere designed by an intelligent Creator. Religious philosophers were also quick to point out that the logical consequence of a godless materialistic universe is ultimately mass murder. One only needs to examine the beliefs and behaviors of Stalin, Mussolini, Hitler, and just about every other murderous dictator of the 20th century to see atheism in action.

Science responded to this by pointing out that belief in the supernatural has led to embarrassing incidents like the Crusades and the Salem Witch Trials. They started applying Darwinian philosophy to social development and, in what amounts to a hostile takeover of firmly held religious ground, claimed to explain morality, ethics, good, evil, right, and wrong purely in terms of biology and chemistry. In fact, they even went so far as to claim discovery of the evolutionary mechanism that predisposes humans to a self-delusional belief in deities!

So where has all of this nonsense left us? Belief in evolution is now equated with apostasy. Belief in God is tantamount to a mental disorder. Is there any way out of this mess? I believe that there is, but it has nothing to do with new scientific discoveries that can debunk evolutionary claims or fancy interpretations of Scripture that somehow accommodate them. Those things only sink us deeper into the holes we've dug for ourselves.

Science, Society and the Church

How is the average person supposed to figure this all out? Most adults, unless they have a technically oriented career or just love to read scientific literature, have forgotten most of what they learned in high school about biology, chemistry, and physics. Lacking a basic understanding of fundamental scientific concepts, they probably aren't very likely to keep up with the latest discoveries, developments or scientific theories. Who has time for that?

Another problem is that most of what's available on these subjects is fairly technical in nature—at least more advanced than what the average person can tolerate for leisure reading by the pool. Even the books supposedly written for consumption by the general public still require a good foundation of scientific understanding. And as interesting as these books may be, they generally lack a broad appeal because they fail to connect with the average reader on a personal level before overwhelming them with technical jargon.

At the risk of offending my primary audience, I'll go out on a limb and say that modern evangelical Christians are probably less scientifically sophisticated than the general American public. It's not a question of intelligence or learning capacity; it's a cultural problem. There is definitely an underlying skepticism of modern scientific theories within the church. But when one considers that atheists, agnostics, and liberal Christians have used science to undermine conservative religious beliefs, such skepticism is understandable. Consequently, those of us who retain an interest in science most often read literature written from a distinctly Christian perspective. Our tendency is to isolate and protect ourselves from the things we perceive as threats to our faith. We've actually become quite proficient at this sort of thing. We have *Christian* radio with *Christian* music, *Christian* television with *Christian* programming, and *Christian* cartoons for the kids to watch after they get home from their *Christian* schools, unless of course it is summer, and they're at a *Christian* camp.

Obviously, there is nothing inherently wrong with any of these things. So why not have a distinctly *Christian* version of science as well? We can call it "creation science" to distinguish it from the evolutionary versions. In our shortsightedness, we tend to treat science like we treat entertainment in that we don't want to support anything that offends our moral sensibilities. Like replacing the profane language of a secular movie with harmless utterances, we try to make mainstream science "safe" for Christian audiences by substituting creationism for evolutionism.

Along these lines, why not have a "science rating system" such as we have for movies, television and video games? We could assign a rating of "YE" for *Young-Earth* creationism, "ID" for *Intelligent Design*, "OE" for *Old-Earth* creationism, "TE" for *Theistic Evolution* and lastly, the most dangerous of all scientific ratings, the dreaded "D" for *Darwinian*. That way, we could know exactly what we're about to get before we tune in. No longer would descent Christian families be caught off guard during prime time by scientific profanities like "billions and billions" from the likes of Carl Sagan. All we'd have to do is look at the rating beforehand. In fact, we could even program our cable boxes so that NOVA, PBS, and the Discovery Channel can only show D-rated shows after 9:00 pm, if ever! After all, we can't allow our children to be corrupted by dangerous ideas that ultimately lead to apostasy, now can we?

On a more serious note, some believers do lose their faith by studying the natural sciences. Thankfully none that I've known personally, but I have read accounts of good Christian kids going off to college to study biology or astronomy only to have their entire world come crashing down on them when what they learn in the secular classroom is at odds with what they have learned growing up in church.[5] This is by far the most costly tragedy of the creation/evolution controversy.

Are the people who leave the church over this debate just collateral damage in a necessary war over the authority and infallibility of the Bible? If they are casualties of war, then I would also have to consider them victims of friendly fire, injured by the ultimatums that other Christians tend to impose on them. How many times have you heard it said that if every last physical detail of the Genesis creation account is not 100% scientifically and historically factual as determined by modern empirical and forensic analysis, then our entire faith is meaningless? That is certainly one way to look at it, but does a firm commitment to biblical inerrancy and infallibility *require* this kind of ultimatum? I'm going to try to convince you otherwise.

We may not realize this, or like to acknowledge it, but modern Christians are infected with the same materialistic philosophy that dominates secular thinking. We have bought the post-Enlightenment lie that *truth* is most accurately expressed in rational scientific terms. So naturally, we feel compelled to comb through the ancient Scriptures, meticulously looking for any text that can be manipulated to settle contemporary scientific issues. To us, this is how we provide the world with uncontested proof that the Bible is still relevant.

---

[5] Nancy Pearcey, *Total Truth: Liberating Christianity from its Cultural Captivity* (Wheaton, IL; Crossway, 2004), pp. 19, 223-224.

But when skeptics don't find the evidence compelling, they simply respond to our challenge in the negative by concluding that indeed, our entire faith must be meaningless—just as we said.

We mistakenly think that such all-or-nothing claims are necessary to defend the Scriptures against scientific challenges. The sad irony is that these statements only undermine the transcendent status of biblical truth. In our zeal to ensure that the Bible remains relevant today, we are actually subverting its timeless message. Like a short sighted person who takes his towel with him into the pool so he'll have it handy when it's time to dry off, the tactics we use to defend the God-given Scriptures against secular claims can unknowingly compromise the very authority that we are trying so desperately to uphold.[6] I'm not saying that the war is over, nor am I calling for a cease-fire; but it is certainly time to reevaluate our tactics.

A Personal Journey

I'll never forget the first book on creation given to me as a young teenager.[7] In those pages, I faced two opposing ideas: either the entire universe was created by God in a span of six days less than 10,000 years ago, or it had arisen by itself over billions of years through blind, purposeless and meaningless forces. Given a choice between those two alternatives, the issue was easily settled. What place could a godless idea like *evolution* have in a biblical worldview?

Nevertheless, I remained fascinated by the world of science, particularly astronomy and physics. Grasping a new scientific concept for the first time is like unlocking the hidden mysteries of the universe. Mysteries that were once known only to God, but are now accessible to man. The fact that we can even do this is quite amazing if you think about it. On the physical level, we humans are built from the same basic stuff as the rest of creation, literally the "dust of the earth." We are products of and subject to the very same physical laws and processes that seem to govern the rest of the universe from the smallest quark to the largest galactic supercluster.[8] So in a very real sense we are intimately connected to the rest of creation. Yet because we bear God's image, we can pattern our thoughts after His thoughts, exercising a measure of conscious autonomy that enables us to ponder our own existence and harness

---

[6] This analogy is from John H. Walton, *The NIV Application Commentary: Genesis* (Grand Rapids, MI; Zondervan, 2001).

[7] Josh McDowell, *Evidence that Demands a Verdict* (San Bernardino, CA; Here's Life Publishers, 1972).

[8] A *quark* is one of the two most fundamental particles of matter. A *supercluster* is the largest structure that has been confirmed in the universe. It is literally a cluster of galaxy clusters.

these very same physical laws for our betterment—or as is often sadly the case, our destruction.

Maintaining a healthy interest in the natural sciences while growing up in the church, I was fascinated by the idea that clues to understanding the mysteries of creation can be found in the book of Genesis. It was also exciting to think that new scientific discoveries could prove God's existence, the truth of His Word, and that He created the universe according to the Scriptures. What a powerful statement Christians could make by showing the world that the claims of the Bible rest on solid science. Unbelievers would be left with no choice but to submit to the overwhelming evidence for the biblical creation account!

Like many of you, I was motivated to find out exactly how the Genesis creation account squares with the latest scientific theories. Wanting to be faithful to the teaching of Scripture, I fully embraced the *Creation Science* movement as the only biblical approach to understanding natural history. I spent many hours reading Creationist literature and watching Creationist videos. But unlike most passive observers in the movement, I spent a little time studying the other guys as well. After all, it helps to know what the opposition is up to.

Over time, I became increasingly uncomfortable with the Creationist claim that any attack on the scientific integrity of Genesis is also an attack on the Gospel of Jesus Christ. Except for a few obnoxious scientists who seem more interested in forcing their own atheist beliefs down the publics' throat than in actually pursuing scientific truth, the majority of those in the scientific community came across as entirely oblivious to the spiritual implications of their work. So whenever I would read statements like that by fellow Christians, I sensed that somewhere along the way a line had been crossed.

While these kinds of bold proclamations are intended to be a motivational charge to the *faithful*, they often become an irresistible challenge to the *faithless*, unintentionally poking the secular hornets' nest and inviting the world to pick apart the Christian faith. When the claims of Creation Science inevitably find themselves at odds with the claims of secular science, Christian evangelism is reduced to highly technical discussions of astronomy, cosmology, geology, biology, paleontology, and anthropology. To me, all of this seems like nothing but a giant evangelical distraction, clouding the simple Good News message of Jesus.

I also became increasingly unsatisfied with some of the creative ways Christians reinterpret the Bible to agree with the latest findings of modern science. Something just doesn't seem right about putting a modern scientific spin on a book that was originally written for a non-scientific audience. If the ancient Scriptures are supposed to contain timeless truth for all generations, which I believe they do, what could they possibly have to do with our 21st century version of science—which is clearly different from the 18th century version of science, which in turn looked nothing like the kind of science practiced in Medieval times? Will Christians always be playing this game of "catch-up" with whatever the current scientific establishment happens to say? That can't be right.

Reinterpreting the Bible to agree with contemporary scientific theories also sets a dangerous hermeneutical[9] precedent. If we can make Genesis scientifically acceptable by dismissing certain elements of the Hebrew cosmos as merely *poetic, figurative, allegorical, metaphorical,* or *phenomenological,* then what about the biblical miracles, the virgin birth of Christ or His resurrection and ascension? Are these things to be understood as literal events or are they merely representative of something else as well? And who has the authority to decide this?

Playing these literary games with difficult texts doesn't sit right with my Reformed Presbyterian background, where biblical authority is not to be tossed around so casually. The way I see it, once we start asking these kinds of questions, we have already replaced the authority of the biblical *author* with the authority of the biblical *interpreter,* and the author's concerns with the interpreter's concerns. So if we're going to do this, then we had better know exactly where to draw the line, or nothing in the Bible will be safe from the critical eye of modern scientific analysis and literary criticism.

Over the years, all of these thoughts and emotions have contributed to where I stand today (April 2007). With more of my life still ahead of me than behind me—Lord willing—I'm under no illusions that everything I now believe will follow me to the grave. And last I checked, publishing a book doesn't waive one's right to change their mind when necessary. In fact, I expect to! Nevertheless, I do feel as though I have reached a rest stop on this journey: a place that affords me a brief moment to capture what I've learned thus far (before I forget what I went through to get here), so that anyone else who also finds themselves unsatisfied with how both sides have framed this debate might profit from my experience.

Historical Perspective

Sometimes by studying the past, we can temporarily detach ourselves from the emotions that our current controversies generate and see how previous generations of believers handled similar controversies that arose in their day. History is full of interesting examples from which we learn. Throughout

---

[9] *Hermeneutics* is the "science" of interpreting the written word, to find what the text actually meant

the book, I will make connections between our struggles and those of the past. In fact, I'll make one right here just to get you thinking about some things.

In the early church, there were philosophical debates about the properties of the four Greek elements: Earth, Fire, Wind, and Water. Some argued that since rocks clearly sank in water, *Earth* was a heavier element than *Water*. That seems to make sense. But others argued from the Scriptures that *Water* was the heavier element despite the ease with which the contrary could be demonstrated.

Some may ask, "What do the Scriptures have to say about the natures of the four Greek elements?" Our modern minds would rightly answer, "Nothing." In fact, such questions don't even come to mind. But don't just take my word for it; consider the wise words of St. Augustine (A.D. 354–430), one of the theological giants of the early church. He responded to these debates with the following statement:

> Let no one think that, because the Psalmist says, "He established the Earth above the water," we must use this testimony of Holy Scripture against these people who engage in learned discussions about the weight of the elements.[10]

The intent behind this statement was probably nothing more than damage control. A few shortsighted Christians must have been making demonstrably false claims based on the Bible, ultimately embarrassing the church. But if we dig a little deeper, we can see how these pointless controversies are inevitable when Christians filter the Bible's content through contemporary scientific paradigms and completely ignore the scientific context of the passage in question.

For instance, the ancient Israelites would not have been too concerned with the four Greek elements and neither are God's people today. Nonetheless, for a brief moment in our scientific history—Earth, Fire, Wind, and Water were all the rage. One might ask, "Since the Bible *is* timeless truth, relevant to all generations, surely it must be addressing a contemporary issue when it speaks of *Earth* and *Water*—right?" But if they were right 1600 years ago, then why don't Christians today still interpret Psalm 133:6 in terms of the relative density of the elements? Have the timeless Scriptures changed their meaning?

The main difference between then and now is that scientists no longer concern themselves with the four original Greek elements. There are now 117

---

[10] St. Augustine, *The Literal Meaning of Genesis*, translated and annotated by John Hammond Taylor, S.J., 2 vols. (New York, NY; Newman Press, 1982), pp. 47-48.

confirmed elements on the standard periodic table, and a host of sub-atomic particles from which even these are constructed. So clearly, 16 centuries of historical perspective can affect what Christians think is important, and how we solve different types of problems.

Accordingly, Christians today don't typically search the Scriptures for facts about chemistry because we now have other useful ways of discovering these things. But modern Christians still tend to read contemporary science into the Scriptures without considering the consequences. How seriously will people take our theological claims about sin and salvation if we continue to make demonstrably false claims based on a misapplication of the Bible? St. Augustine had something to say about that as well:

> They will more readily scorn our sacred books than disavow the knowledge they have acquired by unassailable arguments or proved by the evidence of experience.[11]

Does this sound familiar? Which of our pointless arguments today will future generations of Christians look back on with this same sense of embarrassment?

This isolated example shows us how dragging the book of Genesis through a few thousand years of scientific discovery probably just guarantees that we're going to completely miss the original point of it. And if we really are missing the point of it, then how can we be certain that the issues we spend so much time and energy defending (like the relative weight of *Earth* and *Water*, or the age of the earth) are even important in the overall scheme of things?

The answer to this question is simple, but not obvious. Rather than read Genesis with modern eyes—effectively transporting Moses to our time and forcing him to answer questions that we think are important—we should go back in time, and put ourselves at the foot of Mt. Sinai leaving our modern empirical sensibilities in the 21st century where they belong. How do we go about seeing Genesis through ancient eyes? That's a question I will try to answer.

Why Another Book on Science and Faith?

To be honest with you, there are many other great books that deal with the ideas presented here. Many of them I have read, many I have not—and there are probably several that I have not even heard of. But whenever I find

---

[11] Ibid, pp. 47-48.

myself discussing these things with family, friends or coworkers, I hesitate to recommend any of these other books on the subject. Not because they lack clarity, insight, or scholarship, but because they are simply directed at academic types and primarily focus on academic concerns.

Most Christians who live and work in the "trenches" of the ongoing culture war need something more practical than that, almost like a field manual that gets right down to business without losing the reader in the technical details required by academic scholarship. Now obviously some detail is necessary to make certain points, but I try to avoid anything that doesn't directly apply to the primary arguments being made.

What complicates the study of origins for the layman even more is that most academics are often uncomfortable writing outside of their specific area of expertise. It's actually a legitimate credibility issue that is a big part of academic culture. For example, historians who specialize in ancient Near-Eastern religions are not likely to expound modern scientific concepts. Scientists in turn prefer to remain within their specific disciplines, and the same can be said for the theologians. I'm generally in favor of folks sticking to what they know best, but when it comes to really developing a practical understanding of issues that encompass questions of both science and faith, these artificial barriers that we erect between various academic disciplines make it difficult for the average person to understand how all of these things work together to shape our thinking.

If such a person were to ask me to recommend a "good book" that was short, to-the-point and could help them work through these difficult issues, I would have to give them not one, but several books on a variety of subjects. For example, they would need to know a little about ancient Near-Eastern cultures as well as the history of scientific development. A basic understanding of scientific principles and some familiarity with modern theories including the Big Bang and evolution would also be essential. Finally, they would need to grasp some key theological and philosophical concepts that help tie all of these things together into a single consistent approach. But who has time for all of that? My intention is that this work will fill that void.

Beyond the Firmament is divided into four main parts. Part I asks the question, "What do we know and how do we know it?" Any discussion of science and faith should start here. God reveals Himself to us through both His creation (natural revelation) and His transcendent Word (special revelation). The purpose and limits of these two ways of knowing are explored, and a very practical Christian epistemology[12] is set forth. Some of this material might

---

[12] *Epistemology* is the philosophy of knowledge that attempts to answer the question, "How do we know what we know?"

seem rather basic, but we'll need this before moving on to the rest of the book—so more advanced readers will have to bear with me!

Part II asks the question, "What can the Bible tell us about nature?" In other words, what kind of information about the natural history of the cosmos can we reasonably expect the Bible to provide? My approach here is not unique, but neither is it very common. Rather than force the book of Genesis to answer the kinds of questions that modern, post-Enlightenment, Western Christians tend to ask, this section tries to understand the questions that would have been relevant to a disorganized gaggle of Hebrew slaves that had just been delivered out of the hands of their Egyptian oppressors. Not to make light of the serious problems facing Christians today (like the *life-or-death* question of the earth's age), but what were the specific problems facing Israel as they began their 40-year trek through the wilderness, and how does Genesis specifically address them?

Part III asks the question, "What can nature tell us about itself?" Here we examine the methods scientists use to estimate the age of the earth and cosmos, and how the data are evaluated to piece this information together. There will no doubt be readers who are dismayed to hear just how overwhelming the scientific case is for a universe that is much older than a straightforward biblical exegesis suggests. But after fully absorbing Parts I and II, your reaction to this might not be what you'd expect.

Nevertheless, how should Christians respond to this scandalous news of cosmic antiquity? Should we carefully scan the universe looking for any shred of scientific evidence that can refute these claims? Should we read modern scientific theories into the Genesis creation story and take dangerous hermeneutical liberties with the text—just so the scientists and the theologians can all get along? Should we dismiss entire portions of the Bible as Hebrew fairy tales that can only provide us spiritual guidance and have absolutely nothing to do with the "real" world of facts, evidence and logic? Or should we make a preemptive philosophical strike on the scientific case by insisting that God created the universe to appear billions of years old? For those readers who find these typical approaches wanting, Part III will offer an alternative.

And finally, Part IV is my treatment of the scientific case for evolution. The materialistic philosophies attached to evolutionary thinking obviously have no place in the Christian worldview, and this leaves many believers with no choice but to categorically reject the science of evolution strictly on religious grounds—apart from any meaningful scientific analysis of the subject. But to really grasp the theological consequences of evolution, Christians must first understand the material case for common descent in light of what science *is* and *isn't* able to tell us about the past.

I'm not saying that this is easy. In fact, when it comes to interpreting the evidence for human evolution and migration in light of the special creation of Adam and Eve in the Garden of Eden, there really is no "airtight" solution—not without completely ignoring the testimony of Scripture, the testimony of nature, or both. Anybody who claims to be on top of this situation probably doesn't understand the first thing about either science or theology. For every set of problems solved by one approach, another set of problems arises to take its place. At the end of the day, each individual usually makes up his or her mind based on what set of problems he or she is most comfortable leaving unsolved.

While I believe my treatment of the subject follows naturally from the previous arguments, it admittedly is not a necessary consequence of them. In other words, one can agree wholeheartedly with my understanding of Scripture or the age of the earth all the while disagreeing with my treatment of the biological sciences. Nonetheless, I would be remiss if I failed to demonstrate how the views presented in Parts I, II and III can be applied to this controversial topic. So if you still find yourself unsure of my conclusions, just consider Part IV an informative lesson on how to distinguish the *philosophy* of evolution from the *science* of evolution and how to deal with each on its own terms.

If you are one of those who won't rest until all of Christian theology fits neatly into logical boxes that we humans can wrap our finite minds around, you'll find no repose here. In fact, it might be easier to explain the Holy Trinity ($1+1+1=1$) or the Incarnation ($1+1=1$) then to square the fossil record and molecular genetics with Genesis 1 and 2. But those who can accept the inevitable mysteries that face the mortal mind when it is confronted with the infinite and eternal should not be dismayed by my conclusions.[13]

And finally, I know that many of you like to stuff books like this into predetermined theological categories. I'm sure some have already read the "ending" (whatever that means in a book like this) just to find out where this book falls along the vast spectrum of creation/evolution compromises. "Is the author a Young-Earth Creationist, an Old-Earth Creationist, a Theistic Evolutionist, a Framework guy, an Intelligent Design supporter, or is he even Reformed?" I tend to do the same thing myself. But if I had found any of those approaches to resolving the apparent tension between the Bible and science satisfying, I would probably not have spent the past year working on this book!

If all you do is skim the last chapter or turn right to the final few pages

---

[13] Any honest scientist would say the same thing about his or her discipline with respect to scientific mystery.

searching for some kind of summary, you may or may not find whatever it is you're looking for (hopefully not, since I try to avoid summarizing myself). But how would that profit you? On the other hand, if you actually take the time to consider this material carefully, you WILL grow and be challenged regardless of whether or not your opinions change.

No matter what "side" you find yourself on at the end of the day, there will be consequences. The question you need to ask yourself is this: what are the consequences of your beliefs and can you live with them? Whatever positions you take on issues of science and faith should be thoughtfully considered. In the end, all Christians should share a common concern that our faith remains relevant to culture, that the authority of the Bible stands undiminished, and that the glory of God is not dimmed by our fear of where honest science might lead.

I hope you enjoy the book!

PART I

WHAT DO WE KNOW AND HOW DO WE KNOW IT?

CHAPTER ONE

NATURAL REVELATION

There is so much material out there on the creation/evolution controversy that it's practically impossible for an average person to make sense of it all. At one point, if you were to ask my opinion on the matter, I would have probably just repeated to you the main point from whatever book I last read. Everybody seems to have some convincing arguments. To save time and avoid confusion, many of us gravitate to the stuff that we know is written from a particular point of view. We really don't have time to work through all of the technical details on our own, so we just read or watch things that have already been vetted by someone or some organization that we trust.

If you are like me and you try to read things from all different points of view, you might begin to wonder if all of these people actually live in the same universe. One side "knows" that God created the earth on October 23, 4004 B.C. at 9:00 in the morning. The other side "knows" that the earth is 4.55 billion years old. One side "knows" the fossil record demonstrates that all living things were destroyed in a great flood thousands of years ago. The other side "knows" the fossil record demonstrates that all living things have evolved from ancestral organisms over billions of years. One side "knows" that there is absolute evidence of intelligent design in the universe. The other side "knows" that the laws of nature can produce organized complexity just fine by themselves. The list is endless.

Before picking up another book that claims to "know" this or that, all of this mess should make you stop and think, "How do we really know anything anyway?" The fact that something is true or false shouldn't depend on who is spinning the evidence, should it? Perhaps the confusion has something to do with the possibility that we can know different things by different means? Or could there be limits to the methods we use to acquire knowledge? Most people probably never consider these things, but as it turns out, there is an entire field of philosophy called *epistemology* that deals with what we know and how we know it. Philosophers could write volumes on the subject of epistemology and they have. All I can do is barely scratch the surface, but it's relevant enough to our present dilemma to warrant a brief overview.

Some Things Never Change

How do you know if somebody is telling you the truth about the natural world? For example, I might tell you that I know with absolute certainty that the sun will rise tomorrow at 6:13 in the morning. How do I know this? How can anybody really know that something will happen in the future with such exact precision? Some people charge a lot of money to predict the future and they all have one thing in common: they're all frauds. So how is it that I can tell you, free of charge, that the sun will rise tomorrow morning at 6:13 and be exactly correct? The answer is simple, I have a special phone with a direct line to God and we chat about such things. In fact, I tie up the line so much that the weatherman can't get through, so he has to guess what tomorrow's forecast will be. You don't believe me? I can prove it to you. Go outside tomorrow morning and look to the east. Behold, at exactly 6:13 a.m. you shall see the sunrise![1]

Regardless of your level of scientific education, you should clearly see how unnecessary divine revelation is for predicting a future sunrise. Others may be thinking to themselves, "So that's how they do it!" But a supernatural experience is not always necessary to predict, or "know" the future with some degree of certainty. The reason is so simple that most people don't even stop to think about it. God created the universe in such a way that material things can be described by physical laws. And the same laws that define material cause and effect relationships here on earth are constant throughout the universe and from the beginning of time to the end. This concept is often referred to as the *uniformity of nature* and it works throughout both time and space. Because of this, our universe is intelligible and rational, which is just another way of saying that things are ordered in such a way that we can make sense of them. We can apply what we learn here on Earth to our solar system, our galaxy, all the way to the outermost reaches of the cosmos—backwards and forwards through time.

In the case of the sunrise, it is described precisely by the laws of nature and thus has a perfectly natural explanation. This might seem obvious to us today, but when you consider 6,000 years of recorded human history, looking at the universe as a finely-tuned machine is a relatively modern idea. The ancient Greeks described the sun as a solar disc that was pulled across the sky each day by Apollo's chariot. The ancient Egyptians described the sun as Re, a god who was "born" and "swallowed" by the goddess Nut (the heavens) each morning and evening. Things like this might seem silly to enlightened

---

[1] Of course a 6:13 a.m. sunrise assumes that you are in a specific place on the earth on a specific day. Tomorrow's sunrise at your location will be different, so don't get up and go try it.

moderns such as us, but before the rise of *science*, *mythology* was the medium by which people understood, explained, and interpreted the world around them—and there are important differences between the two.

While *scientific* descriptions explain the functions of the cosmos in terms of material structure, *mythological* descriptions relate the functions of the cosmos to transcendent purpose. Do you think the Greeks really believed in a literal chariot that raced across the heavens each day, dragging the solar disc behind it? Did the Egyptians actually think that Re sailed a literal solar barque through an ocean above the sky? Of course not! Providing the level of technical detail that we moderns demand was not the purpose of ancient mythology.

The point here is that ancient cultures were more concerned with assigning transcendent purpose to the cosmos, and they accomplished this by associating its known functional elements with deities. Even though some of these cultures were highly advanced in terms of having accurate calendars to predict the phases of the moon and other astronomical occurrences, they rarely attributed the causes of these events to a material mechanism. To them, the heavenly bodies were not considered material entities to be understood, but divine beings to be worshiped.

Science as we know it today could not exist in a world governed by a multitude of competing deities with competing interests. Such a universe as this would be an irrational and unintelligible place, completely incompatible with scientific inquiry. So the principle of the uniformity of nature is the foundation of the scientific method.[2] Without it, we have no guarantee that the future will conform to the past, or that the laws of nature in Europe act the same way as the laws of nature in America.

However, the uniformity of nature is actually just an assumption (a very good and very necessary assumption). There is really no way to conclusively prove that all of the laws of nature observed in a laboratory here on earth are the same as all of the laws of nature at work in every other galaxy. There is also no way to absolutely prove that all the laws of nature we observe today were the same ones that have existed all the way back to the beginning of time. And neither can we prove that they will continue to be the same in the future. The only way to know these things with logical certainty would be to observe every particle of matter in the universe throughout all of time. Only then could we really know, without a doubt, that the principle of the uniformity of nature is true. Because we can't actually do that, we have to take what we do know from the little piece of space and time that we have observed and

---

[2] When I use the term scientific method, I'm not implying that all of science can be defined by a single process or principle. In fact there are many different uses of *adduction, deduction, induction* that can all be loosely referred to as the scientific method.

extrapolate these observations to the rest of the universe, past and future.

This seems to work pretty well. In fact, it works so well that most of us take it for granted. For instance, if somebody told you that they saw water flowing *up* a waterfall in Brazil with their own eyes, you would probably be skeptical even though you have not been to Brazil to personally verify that water properly flows *down* every single waterfall in that part of the world. If they showed you a video of this waterfall, you would probably conclude that the tape was playing in reverse. In fact, the principle of the uniformity of nature is so ingrained in our understanding of the material world, that even if you saw a "reverse waterfall" in person, you would still attempt to find a rational explanation.

But what if this person claimed that it was a miracle? Would that automatically end the search for a material explanation? Perhaps for some superstitious Christians who are inclined to see the likeness of Jesus on a piece of burned toast, but most of us would still demand some kind of material investigation. Not because we don't believe in miracles, and not because God is powerless to create such a phenomenon, but simply because there is nothing in our understanding of science or theology that necessitates the existence of a backwards waterfall. It's just too weird!

Now compare our hypothetical waterfall to those magic shows for kids. Most grown-ups spend the entire show trying to figure out how the tricks work. Why? Because we "know" that there has got to be a logical explanation. Our experience in the natural world informs us of how things are supposed to work and these so-called magic tricks contradict that experience. Therefore, we assume that they are not real despite all outward appearances to the contrary. The fact that we sometimes can't even trust what we see with our own eyes when it doesn't fit into our experience demonstrates the power of scientific thinking. We would rather place our trust in an assumption that can't even be proven over something that we can directly observe with our own senses. That's pretty amazing if you think about it. The uniformity of nature may be just an unproven assumption, but if the scientific method is to be of any use to us, we have to start somewhere.

Now some of you are probably thinking to yourselves, "great, I got it, let's move on." But does anybody else find it interesting that all of what we know from scientific inquiry is based on an assumption that can't even be proven with absolute certainty? For all of its appeal to rigorous proofs based on direct observation, science starts with the assumption that the world is rational and intelligible, only to conclude that the world is rational and intelligible. In logic, this fallacy is called *begging the question*. But the good thing for us is that begging the question is the one logical fallacy that doesn't render an argument invalid. It may not provide any new information because you end up right where you started, but neither does it contradict itself.

We know that the principle of the uniformity of nature must be true because it works every time we use it even though it's a circular argument and can't be proven by logical deduction.[3] It is a hypothesis that is always confirmed · hen assumed, so we accept it as a valid theory of how the universe works. It is a *self-validating* truth. It can also be called a *foundational* truth. We believe it is true because if it were not true, the world would not make sense, and nothing else in science could be true. Another way to put it is that we accept the uniformity of nature, not because we can absolutely prove it by deductive reasoning, but because of the *impossibility of the contrary*. In other words, the world just wouldn't make sense without it. So even though science may just be an "empirical faith," it still has the power to describe and define the universe with amazing results.

Keep It Simple, Stupid

Despite the universality of the laws of nature and our amazing ability to piece together accurate descriptions of natural phenomena, not every cause-and-effect explanation of an observable occurrence is necessarily true. A hypothesis is not automatically correct just because it might be consistent with the known laws of nature. There may be many different ways to logically account for the same natural phenomenon. For example, if I go to bed tonight and there is no snow on the ground, and I wake up tomorrow morning to 18" of white powder, I might claim to "know" that is snowed last night, while I was asleep of course.

But did it really snow? How can I be sure? I didn't actually see it snow, but there is this snow on the ground that wasn't there last night. A good lawyer would say that the existence of the snow on the ground is just circumstantial evidence and proves nothing. So perhaps it didn't snow after all. Then how did all of this snow get here? How can anybody really know if it snowed? I'll call my neighbor and see if he knows anything.

So I call my neighbor. He says that somebody came through the neighborhood with a snow machine, just like they do on the Appalachian ski slopes, and spread snow everywhere, while we were asleep of course. He "knows" this to be true.

I ask him if he actually saw the snow machine, and he answers no. So I ask how he can be sure that's what happened? He then asks me if I actually saw the snow fall from the sky.

Touché.

---

[3] Deductive reasoning starts with a premise that is true, and draws a non-trivial conclusion that must also be true.

But there sure is a lot of snow out there, more than what a snow machine could probably make in one night.

Perhaps whoever did this used more than one? That much snow must have required several snow machines.

But where are they?

Someone must have driven them away after making the snow.

But there aren't any tracks in the snow, and moving a bunch of snow machines would leave some kind of tracks.

Could they have been flying snow machines? Certainly that's possible. If we can put a man on the moon, we should be able to make snow machines that fly.

Either of these scenarios is *possible*, but which hypothesis is more *probable*: that it snowed last night, or that somebody has built a bunch of flying snow machines just to make people think that it snowed last night? What kind of crazy person would go to such great lengths just to mislead people? Only someone with too much time on their hands, too much money and an unusual sense of humor would do something like this. Neither of these explanations violates any laws of nature and each one could happen under the right conditions. They are two different analyses of the same data that draw two entirely different conclusions. So which one is more likely to be true?

If you haven't heard of it already, it's time to introduce you to a principle called *Occam's razor*. Occam's razor states that any good explanation of the facts should make as few assumptions as possible, and allow no assumptions that are irrelevant to the question being investigated. In other words, when deciding between two different explanations of the same data, you're better off using the simpler, more direct analysis that assumes as little as possible.

So neither I nor my neighbor actually saw anything happen overnight. My theory is that it snowed while we were sleeping, just like it always does in the winter.[4] His theory is that the snow was delivered by a fleet of levitating snow machines commanded by a delusional unemployed millionaire with nothing better to do at night than trick rational people into making them think it snowed. Who is right? Can I prove with absolute certainty that I am right and my neighbor is wrong?

In science, we seldom have the luxury of *absolute proof.* Rarely can we directly observe the phenomena in question. Moreover, science itself is based on certain assumptions about the physical universe. But most rational people, using principles such as the uniformity of nature and Occam's razor, even if they don't know them by name, would be willing to bet their life-savings that my explanation is correct!

---

[4] If this event had taken place in the summer, then my neighbor might have a better argument.

Common Misconceptions about Science

Science is commonly misunderstood by non-scientists. Unfortunately, these misunderstandings usually end up giving science special powers that it doesn't really have. Sometimes people of faith react negatively to science because we've allowed others to turn it into something that competes with our theological beliefs. But if we really understood what science is and what it isn't, we probably wouldn't cringe every time science rolls out a new theory that attempts to explain something that was once attributed to God.

For starters, science is a tentative enterprise. Even the best theories can eventually be proven false. History shows us how sometimes the most widely accepted ideas can come crumbling down when challenged by new evidence. Science is not absolute truth that must be accepted without question. Scientific theories are just human constructions that attempt to explain the observable facts without using any unnecessary assumptions. As such, scientific theories are always subject to refinement and alteration. If they seem to work, then we keep them around. If they fail to sufficiently explain the observable facts, then we throw them out and make new theories. That's really all there is to it.

We also mistakenly tend to give the laws of nature the power to govern the universe. Laws can do no such thing. A physical law is just a mathematical *description* of how something in nature behaves. It doesn't address the *cause* of the behavior, it just *defines* it. For instance, the law of gravity just describes how objects accelerate toward one another based on their combined mass and spatial separation. It doesn't explain what *causes* the attraction. I don't think anybody knows for certain what causes gravity, but that doesn't mean we can't accurately predict it with a few simple laws.

This may seem obvious, but the existence of gravity does not depend on our ability to fully explain it in terms of a material mechanism. You can still walk outside without any fear of floating off into space. Gravity is a useful physical model that can make accurate predictions, even though the mechanism that causes it is completely unknown. However, when science constructs a law that successfully predicts the behavior of something material, it is easy to start thinking of the law itself as the cause of the material behavior. We have to keep reminding ourselves that the laws of nature are just descriptions; they are not absolute. They are only useful as long as they continue to work.

The example of gravity also demonstrates the difference between a *fact* and a *theory*. On one hand, gravity can be considered a fact. If you drop an apple, it falls toward the earth every time. Since this event can be observed and predicted with significant accuracy, we accept it as a scientific fact. But how it actually works is just a scientific theory. Is there a graviton particle field emanating from the core of the earth that attracts the apple? Does the earth bend the space around the apple and alter its motion? Or does the vacuum of empty space, shielded on one side by the mass of the earth, push the apple down towards the earth? Nobody knows for sure. These are just various theories about the *cause* of gravity.

In summary, gravity is *both* a fact and a theory. This is an important concept because many people will dismiss certain ideas outright because they are "just" theories, even though the idea itself is strongly supported by the facts. Being able to distinguish between fact and theory is important when we evaluate different scientific ideas.[5]

Science Has Its Limits

So far, I've said nothing that should cause anyone to panic. Most people, whether secular or religious, already see the world as an intelligible place where there is uniformity and order. A rational world is simply one that makes sense, where reason and logic have a coherent basis on which to operate. As we have already seen, that basis is the principle of the uniformity of nature. It explains why we can use the laws of nature to navigate our way through the physical landscape.

This approach to the natural world, or nature, is sometimes referred to as *scientific naturalism* or the *scientific method*.[6] The scientific method is the systematic process by which we know material things about the physical universe. It includes things like collecting data, making hypotheses, and testing these ideas by experimentation. The scientific method gives us universally accepted laws that can be used to predict or explain the cause-and-effect relationships of nature. These are the same tools that took the human race from the horse and buggy to the moon in 60 short years.

One thing about science that understandably makes Christians uncomfortable is that it is completely oblivious to anything supernatural. Science is only concerned with physical stuff like matter, motion, and energy. It doesn't assume the existence or nonexistence of anything outside of nature. For this reason, naturalism sometimes gets a bad rap in Christian circles because it is often confused with something called *materialism*. But there are significant differences between the two. Materialism is a belief that *only* material stuff is

---

[5] For more on this, see Moti Ben-Ari, *Just a Theory: Exploring the Nature of Science* (Amherst, NY; Prometheus Books, 2005).

[6] The term *scientific method* is sometimes stretched to cover things that are not necessarily material, like the social sciences. When I talk about science, I usually mean the natural sciences like biology, chemistry, physics, etc.

real. According to materialism, anything supernatural or immaterial is just imaginary and has no meaning.

So it shouldn't surprise Christians when scientific naturalism gets used by atheists to try to disprove the existence of anything supernatural, like God, miracles, prayer, divine revelation and faith. This is unfortunate, and is one of the contributing factors to the suspicion of secular science by people of faith, but science has no jurisdiction outside the natural realm. If I make a theological claim like "God is all-knowing," the only legitimate response from science should be "no comment." Anything beyond that is a stretch for science because a statement like this can't be evaluated scientifically. Of course individual scientists are free to wax philosophical on any topic they wish, including Christian theology, but they cross the line when they invoke scientific naturalism to argue for or against the supernatural.

When it comes to religion, scientific naturalism is not a supporter and neither is it a detractor. It is what it is, a powerful but limited enterprise. I hesitate to call it religiously neutral, but there is a sense in which the *practice* of pure science doesn't require any specific religious or philosophical commitments.[7] Science itself is just a methodology or a procedure, like the process of baking a cake.[8] A cake made by Charles Darwin or Carl Sagan will taste just like a cake made by John Calvin, the Pope, Gandhi, or Mohammed, provided they all use the same recipe. By the way, that would make a great episode on *The Iron Chef*! So as long as we don't expect science to answer religious questions, the philosophical motivations behind it shouldn't affect the outcome if the rules are followed.

That statement will make some readers uncomfortable, but let me put it another way. Let's say you have a neurological disorder and require brain surgery. The procedure that can save your life is very risky. Your chances of survival are 50/50. Your HMO lets you chose between two doctors: a relatively young doctor who has never performed the procedure before and a world-renowned expert with hundreds of successful operations to his credit. Obviously, you go with the more experienced doctor, right?

Now let's say that the more experienced doctor is a militant Darwinist who strongly believes that the human brain is merely the product of chance plus natural selection over time. When he's not performing surgery, his pioneering work in neurobiology is to prove that all of our thoughts and emotions are merely the result of chemistry and biology.

Contrast this with the sincere Christian faith of the young doctor. He is a

---

[7] The only exception I would make here would be to argue that non-Christian presuppositions tend to undermine the uniformity of nature while a monotheistic universe is a more compatible with it.

[8] I have to credit Howard J. Van Till with the "baking a cake" idea.

member in good standing of his local church. He believes that God created
the world in six 24-hour days. He believes that man has a soul and that the human brain was intelligently designed by a sovereign creator.

Now who do you go with? Does knowing the philosophical commitment
of each doctor change anything? Would you rather have the seasoned expert
or the young novice cutting into your skull? Are you more concerned about
the procedure or about the religious beliefs of the one performing it? I think
most Christians would still go with the more experienced doctor.

Now let's turn the tables. Suppose after the surgery, you fully recover.
Praise God for modern medicine! For some reason, the first thing you want
to do is have a deep philosophical discussion about the mind/body duality and
how it relates to the afterlife. Who would you rather have this discussion
with? Are you going to choose the Darwinist whose ideas are guided by an
unyielding commitment to materialism? Or will you now choose the Christian
doctor who will approach the subject from a biblical perspective?

For this, I'm pretty confident that most Christians would choose the
younger doctor. The reason is simple, we direct questions of material procedure (brain surgery) to the most qualified scientists regardless of their religious or philosophical commitment, and we direct questions of transcendent
purpose (mind/body duality) to the person that shares our religious presuppositions regardless of their scientific credentials. So when science is done
properly, Christians should be able to accept the results even if we don't share
the philosophy of the ones drawing the conclusions.

Many of us don't like the fact that science is oblivious to supernatural
things. However, when looking for a natural cause-and-effect relationship,
any supernatural assumptions are "trimmed" away by Occam's razor as unnecessary to the material nature of the question being asked.[9] We may not like
these rules, but that's just the way science works. If we want to ask questions
of an immaterial nature, then we have to step outside the scientific method
and use some of our other ways of knowing, like *philosophy* for instance.
Rather than feel excluded from the marketplace of scientific ideas by unfair
secular bias, Christians need to learn to work within the system by understanding the rules and using them to our advantage.

We have to remind ourselves that science is an inherently limited enterprise. If the scientific method were allowed to (1) give supernatural explanations of or (2) assign transcendent purpose to natural events, it would actually

---

[9] The Intelligent Design movement makes a good argument for using the scientific method to recognize, but not speculate on the nature of, supernatural causality. This appears to be a rational, albeit
controversial, exception to the naturalistic tradition of the sciences and will be looked at in greater
detail in Part IV of the book.

look more like ancient mythology and—as a result—be completely useless to us. Therefore, we intentionally limit science by restricting it to natural cause-and-effect relationships that can be explained in terms of material mechanisms. We have other ways to evaluate supernatural or transcendent claims, like philosophy and theology—or mythology (for those who lived in ancient times).

We also have to realize that since science doesn't have the ability to prove or disprove supernatural claims, they can not be *scientific*, by definition. Now don't misunderstand this point, this doesn't mean that supernatural claims are automatically *wrong*. It just means that science can't make that determination. If your initial reaction to this statement is to take offense, then perhaps you have unknowingly bought the modern lie that only science is able to reveal absolute truth.

The fact that supernatural claims are not scientific shouldn't offend us. And we shouldn't think for a minute that this is some kind of atheistic conspiracy against our faith, or an attempt to keep Christianity out of the public schools. This is just the end of the road for science, and we should be thankful for that because it is our greatest defense against materialistic attacks. Transcendent realities need to be evaluated by philosophy and theology, not science. When John tells believers to "Test the spirits,"[10] he's not talking about laboratory experiments with control groups. The last thing Christians need is to have a bunch of scientists trying to solve the rich mysteries of the faith with test tubes and Bunsen burners.

Science can also be limited by technology because we don't always have the ability to directly observe natural phenomena. For example, we can't send a spaceship to the sun and retrieve a sample to study in a laboratory here on Earth because it would obviously be incinerated. Fortunately for science, there are ways of knowing other than by direct observation. We can still know how the sun shines by inferring this mechanism from other things that we do know, like spectroscopy[11] and nuclear physics. But sometimes, despite the best efforts of scientists and engineers, we just don't have the technology or the ability to figure things out. As a result, there will always be plenty of natural mysteries for future generations to uncover. But if we stay true to the scientific method, we will leave these mysteries unsolved until technology provides a way to solve them.

We should recognize that science has limits and not try to push it beyond those limits. Unfortunately, many well intentioned Christians fall into the trap of arguing that these natural mysteries are absolute proof of God and creation. Now obviously God created all that is, was, and ever will be; that's a given for Christians. But the only thing these natural mysteries prove to us is that sci-

---

[10] I John 4:1

[11] Spectrography is the study of certain interactions between light and matter. The existence of certain elements can be determined by absorption or emission at certain frequencies.

ence can't always figure everything out. That's it folks. It is a big mistake to plug God into the holes left by science. We can be certain that someday these holes will be filled with natural explanations. Where then does that leave us? Furthermore, if we force the scientific method into supernatural territory, the secular scientists will use it as a club to beat our faith out of us. Christians should avoid giving science that kind of power.

Ignorance and Apologetics

The way that Christians tend to deal with the mysteries of creation needs to be addressed. Our intentions are good, but sometimes we don't think about the unintended consequences of what we do or say. Sometimes others do not perceive the arguments we give in favor of God and creation in the same way that we intend them. We need to be open to the possibility that perhaps there are some bad habits that need to be broken.

Here is a perfect example: most of the counterproductive arguments that we use to prove that God created the universe unfairly exploit the inherent limits of science in one way or another. Rather than appreciate and respect these limits, we tend to unfairly manipulate them to our advantage. Now I do not doubt anybody's motives, I'm just saying that we need to be more aware of what we are doing. The unintended consequence of stumbling onto the scientific playing field is that our claims of faith become subject to falsification just like any other scientific claims. Once we lure science out of its cage to play, it's hard to get it back in.

Let me explain how this works. Because science can't explain everything, it's tempting for Christians to capitalize on the limits of science and offer supernatural explanations that support whatever specific version of creationism that they believe in. These arguments may be effective in the short term, but they only work as long as the thing being argued isn't yet established by science. It goes something like this: science can't explain X. We know that the Bible says Y, and Y seems to explain X. Therefore the Bible is true and all men are liars. In logic, that is called an argument from ignorance. It only works because of what we don't know, not necessarily because of what we do know. Something is assumed to be true only because nobody yet knows how to prove it false.

Here is another one: X is something beautiful or amazing, X is so incredible that it couldn't possibly have a natural explanation, the Bible says God established X by the work of His hands, therefore God must have done X by divine fiat, there is simply no other reasonable explanation. In logic, that is called an argument from incredulity. It only works when we just can't conceive of how something could be explained naturally, so we say that only God could have done it.

There are many problems with these two lines of reasoning, which are really two sides of the same coin. But the biggest problem is that we unintentionally put matters of faith in opposition to scientific discovery. What happens when scientists eventually find a natural explanation for *X*? Does God then become unnecessary? Does Jehovah join the ranks of the unemployed deities from the ancient world because we no longer need Him to explain the mysteries of creation?

Christians have been doing this kind thing for centuries. For example, before the Italian astronomer Galileo Galilei (1564–1642) provided conclusive evidence that the earth revolves around the sun, it was common knowledge that the earth was fixed at the center of the universe and the entire cosmos revolved around it. One of the unsolved mysteries of the Middle Ages was how the earth could just hang there in midair while the entire heavens rotated around it daily. Like all Christians who lived before Galileo, the great Protestant Reformer John Calvin (1509–1564) naturally filtered the Bible through his own geocentric[12] understanding of the universe. Commenting on Psalm 93:1—which says, "Indeed, the world is firmly established, it cannot be moved."—Calvin wrote:

> How could the earth hang suspended in the air were it not upheld by God's hand? By what means could it maintain itself unmoved, while the heavens above are in constant rapid motion, did not its Divine Maker fix and establish it?[13]

I'm generally a big fan of John Calvin, but this statement is extremely short-sighted because it fails to consider the progressive nature of science.

Less than 100 years after Calvin made these remarks, Galileo discovered that the earth actually orbits the sun; and before another century would pass, Isaac Newton (1643–1727) had discovered that the earth is actually held in its orbit by the force of gravity—as are all of the other planets. Based on Calvin's explanation of how the geocentric model depends on God's providence, these scientific discoveries might lead one to wrongly conclude that the universe operates without any need for a Creator. Quite frankly, our medieval apologetics and the scientific revolution that followed could have been what drove many post-Enlightenment Christians to deism[14] and atheism. Could we be doing the same thing today?

Most popular creationist literature is filled with examples of this. Check it out for yourself. Just this week while sitting down to eat I saw my daugh-

---

[12] *Geocentric* simply means earth-centered.

[13] Commentary on Psalms: Volume 4. http://www.ccel.org/ccel/calvin/calcom11.ii.i.html

[14] Deism is the idea that after God created the universe and set everything in motion, He then withdrew any divine influence and allowed creation to operate independently.

ter's fifth grade Christian astronomy book lying on the kitchen table. I decided to flip through it out of curiosity. I turned to Chapter Three, "The Origin of the Universe" and saw the following:

> One of the biggest problems for those believing in cosmic evolution is explaining where all the structure in the universe came from. How could stars form and then organize themselves into galaxies, and how could the galaxies form clusters of galaxies? Scientists who believe in evolution have no answer to this question, because no one has ever seen stars (or anything else) arising out of nothing.[15]

This is a *textbook* case of arguments from ignorance and incredulity (pun intended). First of all, stars don't arise from nothing and I've never heard anybody claim that they do, which probably explains why nobody has ever seen it happen.

Secondly, every astronomer knows that stars are formed when clouds of interstellar dust and gas, mostly hydrogen, fall in on themselves under the influence of gravity and the intense pressure and heat ignites the nuclear furnace that fuses the hydrogen into helium and releases tremendous amounts of energy. Some of the details are still a little sketchy, but no supernatural explanation is necessary for that. The laws of nature are also pretty good at generating complicated structures out of disorder in open systems. Things like crystals, snowflakes, sand dunes, hurricanes and spiral galaxies are all good examples of this.[16] No leap of faith is necessary there either. The mechanisms that do this are all very well documented in the available scientific literature.

But let's assume, for the sake of argument, that we really don't know any of these things. It's possible that this could be a really old astronomy textbook written before our modern telescopes, looking deep into space, had revealed stellar nurseries with hundreds of stars in various stages of life. Perhaps it was

---

[15] Johnathan F. Henry, *The Astronomy Book* (Green Forest, AR; Master Books, 2002), pg. 14.

[16] The 2nd law of thermodynamics states that in a closed system, things will always move from a state of order to disorder, increasing the total entropy of the system. It is frequently misunderstood and misapplied by Christians, almost to the point of embarrassment. It would be impossible to address all of the abuses here, but whenever you see somebody using it just remember this: most things in nature are not closed systems because they are free to interact with their environment. For instance when a large hurricane organizes itself over the Atlantic Ocean, the negative entropy of this highly organized and complex structure is offset by the positive entropy of the ocean, the atmosphere and the coast. When the whole thing reaches equilibrium (after the storm dies out), the end result is always a net increase in entropy (disorder) for the planet. All you have to do is look at whatever coastal town gets demolished to see the net increase of disorder. The point is that the 2nd law doesn't forbid complex structures from organizing on their own, as long as they are offset by a corresponding amount of disorder.

written before high speed supercomputers, able to speed up the clock on the laws of physics, ran simulations showing how newborn stars can organize into galactic structures under the influence of gravity. Or maybe it was written before our largest particle accelerators showed virtual particle/antiparticle pairs "arising out of nothing" before quickly annihilating each other in a burst of light. So I could be gracious and give the author the benefit of the doubt.

But here is the real problem. Should science abandon its naturalistic quest just because the task of figuring something like this out seems too hard? Do we expect astronomers to just throw up their hands and say, "Oh well, I guess God must have done it—let's go home early today?" If not, then what is going to happen when these things eventually get discovered? Where will that leave us?

We already know that science is a limited enterprise, and when it can't tell us everything we need to know about the universe it's all too tempting for Christians to rush in and fill the void with a supernatural explanation. We want to show the world that where science has failed, we have succeeded! To us, this proves conclusively the truth of our supernatural claims. We have the answers if only people will listen!

I know what I'm talking about because I used to be an expert at this. We think we're standing up to secular science and showing the world that the Bible alone answers the mysteries of the physical universe. But I don't think we ever pause to consider how we come across to others. If explaining the unexplainable is all that our God is good for, what need does science have for Him? For those whose profession and purpose in life is to explain what was once unexplainable, we've just made God unnecessary to them. We've just placed a huge stumbling block between them and the Gospel.

Why would we do something like this? Is our ultimate goal just to poke science in the eye whenever we see an opportunity, or is our purpose in life to draw all people to a saving knowledge of our Lord Jesus Christ? No matter how tempting it is to trump science by taking cheap shots at its incomplete knowledge, I think we should avoid using God as a placeholder for the unsolved mysteries of creation.

True to form, the very next paragraph in my daughter's astronomy book says this:

The Word of God, however, does answer this question. The Bible says that God created and organized the universe by His infinitely powerful spoken word.[17]

---

[17] Ibid, pg. 14.

Put away the telescopes folks, now we know how the stars and galaxies are formed! Despite the gratuitous sarcasm, I really don't have any problem with this statement, as long as it isn't offered as a divine placeholder for a yet unknown material mechanism. In fact, I would even make the case that really young children don't need to know anything more about astronomy than "He made the stars also."[18] But we shouldn't misrepresent modern astronomy and the Bible to kids by telling them that these two ideas are both competing answers to the same question.

I agree with Psalm 33:6, which proclaims "By the word of the Lord were the heavens made." But if stars can have a natural cycle of birth, life, and death, why put this material mechanism of creation in competition with the spiritual meaning of creation as revealed by Scripture? If I remember the first question of the *Catechism for Young Children* correctly, didn't God also "make" us? Yet were we all not also born from our mother's wombs as part of a natural cycle of birth, life, and death? What happens when my daughter goes off to college, takes Astronomy 101 and finds out that we've actually known for years the physical mechanisms by which stars and galaxies were formed? Will she then question everything she learned about God? Will she feel betrayed by her parents and teachers?

Science has a very successful track record of discovering things that were previously unknown. If we raise our children to believe that supernatural explanations are in competition with natural ones, we are basically entrusting their salvation to ignorance and incredulity. For their sake, we should always leave the door open for the future discovery of a material mechanism that doesn't threaten the theology of creation that we give them.

I realize that many readers may have picked up this book for the primary purpose of figuring out how to provide their children a scientific education that doesn't water down science or compromise their faith. One of the best things we can do when it comes to educating Christian kids is to avoid all arguments from ignorance and incredulity! For example, our approach to astronomy should be that God created and organized the heavens (primary cause) regardless of whether or not science ever figures out exactly *how* He did it (secondary causes).[19] By using this approach, the search for a material mechanism becomes a challenging quest to reveal the character of God by

---

[18] Genesis 1:16

[19] If you're not familiar with theological terms: "primary causes" and "secondary causes" here is quick introduction. Many times, the Bible speaks of primary causes, such as in Psalm 33:6. However, God also uses secondary causes in order to accomplish the effect of the primary causes. When the Bible says "By the word of the Lord were they heavens made" it doesn't exclude a discernable material mechanism that also served as secondary cause. Secondary causes are basically the laws of nature.

discovering the patterns by which He governs the cosmos, rather than an opportunity to lose one's salvation through scientific discovery.

Out of curiosity, I flipped to the front cover of the same textbook and saw that it was printed in 2002! Either the author doesn't know much about contemporary astronomy, and is therefore unqualified to write a textbook for children, or he is deliberately ignoring widely accepted scientific explanations. Why would a Christian man intentionally misrepresent modern astronomy to children? As I flipped through more of the book the answer became clear. The book really isn't about teaching children astronomy, this book is all about keeping children from discovering the natural explanations of what we see in the heavens.

By forcing our children to choose between modern astronomy and the Bible, we consign them to either perpetual ignorance or eternal damnation. That doesn't sound like much a choice to me. For some reason, we tend to think that if we can just keep our little ones ignorant and incredulous of the natural world, they will not be enticed by naturalistic theories when they leave home or take biology and astronomy in high school. We wrongly assume that they won't have any interest in the material "machinery" behind the spiritual curtain. We don't want their knowledge of God to be supplanted by natural cause and effect. I understand the motivation here, but sooner or later they will look beyond the firmament. And what happens if they realize they have been misled by those to whom their very salvation was entrusted?

If we fail to teach our children how to successfully integrate both the primary and secondary causes of nature into a biblical theology of creation that sees God providentially working through the patterned behavior of ordinary material cause and effect before they leave home for college, they might just reject the primary causes all together and embrace a deistic or atheistic worldview. After all, who needs God if science can explain everything? If you have any children's books like this in your home, I suggest you get rid of them.

For Bible-believing Christians, abandoning the scientific search for supernatural causality may sound like retreat from the world, but this is actually our greatest defense against those who use science as a weapon against our faith. One thing we need to understand is that the door between science and the supernatural swings both ways. Once we open it, we had better be prepared for what comes through. When we undermine the physical cause and effect relationships of scientific naturalism by offering supernatural explanations of natural events, we bring the whole Counsel of God into the realm of scientific inquiry, and the one thing that scientists are really good at is testing hypotheses and falsifying claims. By doing this, we not only violate God's command against putting the Lord to the test, but we abandon our greatest defense against secular science by placing our entire faith under its jurisdiction.[20] Once we give science that kind of power, we have only ourselves to blame when those within the scientific community use it as a weapon against the transcendent realities of the Christian faith.

You may take issue with this approach to science and the supernatural, but isn't this the way we operate each and everyday? If you break your leg, do you not go to the doctor? If he tells you that you actually have an evil spirit living in your leg, would you not leave the office immediately and report him to the proper authorities? If your toilet gets backed up and your plumber starts to dance around the bathroom sprinkling flax seed in some kind of strange ritual to an unknown toilet god, would you not get another plumber? Would you even waste two seconds arguing with him about the uniformity of nature or Occam's razor? If so, you are more patient than I am.

You may think that these two examples are silly. You mean you don't believe in witch doctors and toilet exorcists? That's exactly my point. We expect the world to conform to our everyday experience. But at the same time, we know that there is a supernatural economy that transcends the material world. Our confusion is a result of not properly understanding the capabilities and limitations of each. We may go to the doctor for the broken leg, but don't we also pray for a speedy recovery? Are we speaking from both sides of our mouth? Do our actions betray our faith? I don't think so. The next chapter looks at our knowledge of the supernatural and how this differs from what we can know about the natural world.

---

[20] Deuteronomy 6:16

CHAPTER TWO

SPECIAL REVELATION

Christians place a high value on the Bible as God's inspired Word, which contains the 66 books of the Old and New Testaments. In fact, the Christian religion begins and ends with the Word of God. It is the ultimate authority on all matters on which it speaks. But how do we know the Bible is true? How can we prove that it is the Word of God? Christians have tried all kinds of arguments to prove the truth of Scripture. I could go down the list one by one, but the only one you really need to remember is II Timothy 3:16, "All Scripture is God-breathed and is useful for teaching, rebuking, correcting, and training in righteousness." That's it—next question please.

"But wait a minute" you say, "You can't use the Bible to prove that the Bible is true! That's just begging the question! That's just a circular argument! That's a logical fallacy!" Actually, the Bible, like the principle of the uniformity of nature, is a self-validating truth. It can't be proven by anything else because nothing else has authority over it. Anything used to try to prove the truth of the Bible becomes a higher authority than the Bible. And yes, this even includes the Holy Grail of Christian apologetics: scientific evidence.

Why do we seek validation from science when science itself rests on an assumption that can't be proven? As tempting as it is to point out passages of Scripture that seem consistent with scientific evidence, what we are really saying when we do this is that science is more reliable than the Bible. If we are going to maintain that the Bible is the ultimate authority on all matters to which it speaks, then it does no good to betray its authority by seeking approval from science. Like science, the Christian faith also rests on a self-validating assumption that can't be proven by logical deduction.

While most of us wouldn't think to search the Scriptures for information about calculating the sunrise, determining the composition of stars, mending broken bones, or fixing toilets and automobiles, we do trust God's Word to make us "wise for salvation through faith which is in Christ Jesus."[1] In fact, there are many things that can be known *only* by studying the Scriptures. Because of this, the Bible is often referred to as *special revelation*, as opposed

---

[1] II Timothy 3:15

to *natural revelation* which refers to those things that can be learned from nature. Spiritual knowledge deals with things like the character of God, the condition of man, the sufficiency of Christ, and the Kingdom of Heaven. Since these realities transcend physical space and time, special revelation should never find itself in opposition to natural revelation. Can God contradict Himself?

Even non-religious people claim to know a lot of things that have no basis in the physical universe. Whenever people discuss things like good and evil, right and wrong, love and hate, mercy, justice, beauty and meaning they do so on a totally different level than when they discuss material cause and effect. Some refer to these as *metaphysical* realities as opposed to the physical or material realities of science. Working on this level requires different tools. We put the empirical tools like the scientific method back into the toolbox and pull out things like theology, tradition, and philosophy.

The Materialist Universe

We know that there is more to life than physical matter not because we can prove it by logical deduction, but by the impossibility of the contrary. Materialism, as a philosophy of life, is simply not consistent with the human experience. The spiritual realities that transcend the material universe are necessary because without them, the universe would be a morally unintelligible and irrational place. Immaterial absolutes like good, evil, love, and hate would be entirely unknowable, and would basically be up for grabs. In that kind of universe, anybody could create their own individual morality and it would be just as valid as anybody else's. This idea sounds pretty fair until one person's individual value system infringes on somebody else's individual value system.

For example, if somebody were to say that all persons with blonde hair must be shot, there is no materialist defense against this kind of thing. Any attempt to show that this is wrong or immoral would appeal to things like personal rights, individual liberty, good and evil, tolerance, and acceptance of others. But none of those things have any absolute meaning in a materialist universe. They can mean whatever anybody wants them to mean because they are only products of our thinking, figments of our imagination, which itself is just based on chemistry and biology. Materialism forbids any transcendent realities that exist apart from our own devices. Concepts like right, wrong, good, evil, fairness and justice are just human inventions, derived out of a necessity to maintain some semblance of social order for the *common good*. But ultimately they are whatever we say they are.

If you want to argue that the shooting of all blondes is contrary to good social order, a materialist can just as easily say that he doesn't believe in social order and your definition of good doesn't apply to him. In fact, if he can convince a civil majority to agree that all blondes should be shot, they can just make a law and that would settle it. And if he can't get a popular majority, all he really needs is a powerful minority and they can "lock and load." Who needs a majority anyway? After all, the only legitimate authority in a materialist universe is power. Might makes right. All things would be permissible and all civil, familial, and ecclesiastical authority would be illegitimate. Any appeal to transcendent truth is equivalent to believing in Santa Claus and the Tooth Fairy.

Ironically, materialists today enjoy the relative safety and security of a morally intelligible universe and at the same time speak out so boldly against it. Fortunately for them, they will never have to face the consequences of their own philosophy because most people, whether religious or not, know deep down inside that there *are* moral absolutes. That is the way most people live their lives. This even includes the "academic" materialists. They go about their business in a quiet defiance of their destructive beliefs. They love their families, they maintain the ethical standards of their profession, they submit to the civil authorities, they are outraged by human injustices, they have compassion for others, and they believe that people's lives have meaning and purpose.

What this shows us is that materialism, as an assumption, is blatantly inconsistent with the human experience. Fortunately, most materialists borrow enough theistic ideas to make sense of their atheistic world. So I guess deep down that makes them closet theists. When engaged in discussions with people like this, all you really have to do is point this fact out to them and convince them to "come out of the closet."

The Patterns of Providence

None of this should come as any surprise to the theist who has faith in immaterial absolutes. Nature is not our only source of knowledge. We can clearly know different things by different means. For example, I know that I love my wife and kids. A smart neurobiologist may provide a chemical explanation for the love I feel for them on an emotional level. To that I say, "so what?" I expect nothing less from a person of science. But if materialism is true, then chemistry and biology are all that there is to it. Some may choose that reality because they think it to be intellectually honest, but not many people can consistently live like that. Christians readily accept that there is a spiritual economy that transcends the physical universe. Like science, it is a self-validating reality and cannot be proven logically.

On the metaphysical or spiritual level, I know the love I have for my family is real and has meaning. I know that this love, while imperfect and subject to my selfish tendencies, is real because God's love for us is real. As creatures made in His likeness, we are created with this same capacity to love and our imperfect love for one another is but a shadow of His perfect love for us. Advances in the field of neurobiology have nothing to do with this. I don't care if the emotion of love can be fully explained by chemistry! Does that make it any less real? In fact, I would be surprised if there wasn't an underlying material mechanism from which all of our emotions are felt and experienced by our physical bodies. Why else would we have all of these billions of neurons and chemicals inside our skulls?

Does God not use ordinary means to accomplish His will? If His will is for me to love my family, then shouldn't I at least have the hardware needed to experience it? There is a reason why our brains are larger and more complex than those of other creatures. Why should we be worried when science tries to explain how all of that extra hardware contributes to our conscious activities? Are we worried that our souls will be rationalized away by science? Give me a break. Science can never do that! All it can ever hope to do is reveal the physical mechanisms behind our thoughts and emotions, and even that is a stretch given our current progress.

This interesting relationship between the material world and the spiritual world is best understood in light of the wonderful but mysterious Christian doctrine of providence. Providence tells us that somehow God, in His infinite wisdom and power, accomplishes His will by natural means using secondary causes, respecting the very patterns of nature that He established so that we can live in a coherent universe that is—from our finite perspective—both *necessary* and *contingent*.[2] Without the idea of providence, the uniformity of nature by itself may lead one to accept Enlightenment deism, the belief that everything in the universe runs by itself and God is just a passive observer. But the doctrine of providence mysteriously upholds the natural order of the universe without limiting God's sovereignty over it. Providence allows the primary and secondary causes of nature to coexist in perfect harmony.

It's important to note that this is not an either/or situation. These two ideas are not competing for the same territory. Unfortunately, Christians can tend to lose sight of that. When Isaac Newton first proposed that gravity was the ma-

---

[2] *Contingent* is the opposite of *necessary*. In nature, a necessary event is something that has to happen. If I drop a rock from the roof of my house, its fall to the ground is a *necessary* event. Things that result from the exercise of free will, such as my choice of either chocolate or vanilla are *contingent* events. In a strictly materialistic universe, all events would be necessary and true choice would be just an illusion.

terial mechanism responsible for the motions of all heavenly bodies, the initial response he got from the church is a great example of what can happen when we confuse material and spiritual realities.

Recall that during the Middle Ages, the universe was not seen as some finely-tuned machine operating like precision clockwork in accordance with the immutable laws of nature. That is a fairly modern view of the world. Back then, the doctrine of providence was understood to literally mean that everything in nature was contingent upon God's will, especially the heavenly bodies which were not even seen as ordinary material entities. And we know that God is indeed sovereign over all of creation, so that is not necessarily a wrong idea,[3] but this resulted in a belief that God kept all things in motion by His literal "hand" of providence![4] Not much consideration was given to secondary causes because none had yet been discovered. Some in the church felt that a natural explanation of the solar system via secondary causes would weaken people's faith by taking God out of the picture and making Him unnecessary. After all, what is left for God to do if *gravity* is responsible for everything?

So poor Mr. Newton, a deeply spiritual man, was accused of taking "from God that direct action of his works so constantly ascribed to him in Scripture and [transferring] it to a material mechanism." Newton was also said to have "substituted gravitation for Providence."[5] Are these charges even legitimate? Of course they are not! Gravity and providence are two different answers to two different questions about the same phenomena—one description assigns transcendent purpose (via a theological medium) and the other defines material behavior (via a scientific medium). Why did the medieval church give science the power to rob the natural world of meaning and purpose by allowing these two descriptions to compete with one another? Why do we still do things like this today?

During the scientific revolution, Christian scholars fleshed out the doctrine of primary and secondary causes, giving us a view of divine providence that unifies science and religion in a beautifully consistent theology of creation. Because of their insight, Christians today are able to see God working through the laws of nature and we typically don't fear any discovery of material causality—unless, of course, we start asking about the natural origin of things. Once that happens, we go back on the defensive and accuse science of leaving God out of the picture. We desperately need a

---

[3] Colossians 1:15-18

[4] For example, see Calvin's commentary on Psalm 93:1 from page 29.

[5] Andrew D. White, *A History of the Warfare of Science and Theology in Christendom* (New York; Appleton, 1898), Chapter I.

more consistent approach.

Miracles, Signs, and Wonders

So far, we've only looked at natural phenomena that have rational explanations. But what about supernatural events like miracles? If we accept the uniformity of nature as a binding principle, how do we know that the miracles recorded in the Bible are true? Wouldn't they be serious violations of the uniformity of nature? Don't miracles ruin the foundation of the scientific method by sabotaging the coherence of nature? Those are all great questions, but I think we first have to ask what defines a miracle. It's really not that easy.

One common definition of a miracle is "anything that science can't explain." If that were true then many everyday phenomena would have to be considered miracles since there are still many things that science has yet to understand. Moreover, that would make just about everything in ancient times miraculous since science could explain very little back then. According to this definition, if all of the secrets of the natural world were to be ever figured out by science, there would be no more miracles (not to mention that science would become a boring enterprise).

Others might say that a miracle is a specific violation of the laws of nature for divine purposes. A good example from the Bible would be Jesus turning water into wine or being raised from the dead. I'm pretty confident that most Christians would classify these events as miracles. But does something always need to violate the laws of nature to be considered a miracle?

For example, if I lose my wedding ring at the beach, then go back the following year and find it while I'm "randomly" digging in the sand, many would say that was a miracle. But was it really? Did anything supernatural actually happen? If the laws of nature were not violated in any way, wouldn't that just be an amazing coincidence? Sure, it could merely be a coincidence, but the odds are so great against it that many would still call this a miracle. Then the next question is how improbable does a natural event need to be before it qualifies as a miracle? A thousand to one? A million to one? A billion to one? Any number we pick is completely arbitrary. You can see the problem with trying to define it in terms of probability.

I've also heard people refer to the birth of a newborn baby as "the miracle of life." Is that really a miracle? A baby being born is definitely an incredibly moving experience. But it happens thousands of times each day and except for the Virgin Mary, there is a perfectly natural explanation for how it happens. Again, this is very difficult to define.

Some say that a miracle is *divine intervention*. But to me, this whole idea of divine intervention is no different than Enlightenment deism accented with an occasional act of providence. It implies that God spends most of His time just sitting around watching the universe unfold according to the laws of nature and the choices of His creatures, until He sees something that needs tinkering with. Then, in a supernatural act, God reaches down and makes the necessary adjustments to the machinery of nature by "divine intervention" so that things stay on track. What is God up to during the rest of the "non-intervening" time anyway? I don't even find this term useful.

This whole idea of divine intervention is a perfect example of how Christians have allowed science to corrupt their theology of creation. Despite its law-like behavior, the universe is not self-sufficient. The creation is entirely dependent on the Creator to continuously sustain and uphold it. God's people didn't think of the universe in terms of modern machinery before the time of Galileo and Newton. Back then, few people even considered the notion that the universe could function apart from its Creator. Every single motion of every single entity was directly attributed to the continuous effect of divine providence, not to punctuated acts of divine intervention augmenting the universe's "ability" to govern itself.

I've even heard some theologians compare this type of interventional providence to a form of idolatry.[6] This might sound a little harsh, but it makes sense if you think about it. The term *intervention* implies that something else was governing and sustaining the universe prior to the intervening act by God. So by definition, that other thing governing the universe was acting in God's place. Even if you say that the universe was taking care of itself, then that just means the universe itself is God. The conclusion is inescapable. As wonderful as modern science is at describing the inner workings of the universe and as useful as the laws of nature are to human progress, we shouldn't let them corrupt our theology of creation.

Remember what we learned from Chapter One. The laws of nature are just man-made descriptions of how things behave. They do not actually cause the behavior that they describe. The actual causes behind the laws of nature are speculative at best. So the uniformity of nature still leaves plenty of room for God to interact with His creation. But don't misunderstand the point here. I'm not saying that these mysteries are *scientific proof* for the providence of God; I'm just saying that our knowledge of the material mechanisms behind nature is not in competition with our doctrines of providence. Let's not sacrifice the wonderful and mysterious doctrine of God's providence on the tentative altars of science!

I think our view of miracles needs to reconnect with the doctrine of prov-

---

[6] R. C. Sproul, "Providence, Science and the Sovereignty of God", from *ETERNITY Magazine*, Copyright 1988, Foundation for Christian Living, c/o 1716 Spruce St., Philadelphia, PA 19103.

idence. What if God, by his "hand" of providence, precisely governs every particle of matter in such a way that everything in nature can be consistently described by a universal set of physical laws? By doing this, He guarantees us a rational world, and the uniformity of nature allows us to make sense of it. As conscious beings created in God's image, we can discover these patterns of providence to analyze and quantify the world we inhabit. Being free moral agents with wills to make choices according to our desires, we can interact with the rest of creation and harness its predictability to advance our civilization to the glory of God.

Obviously in such a universe as this, God could at times, in order to display His power, govern not according to the regular patterns of providence, but in an irregular way, bypassing the secondary causes and acting directly through primary causes. For a brief instant, nature would no longer behave as it normally does. Because these rare events are contrary to the uniformity of nature and to our everyday human experience, we call them *miracles*. But these events are ultimately no different than the usual patterns of providence in terms of divine governance. Having water suddenly turn into wine is no more an act of God than fermenting grapes, which take water from the ground, with yeast in jugs over time to make wine according to the "laws" of nature. In both cases, water was "turned into wine" by God.[7] So why do we only call one a miracle? That's a good question. I guess some things just have a way of getting our attention more than other things do!

So if the Christian worldview rests on the theology of creation as its epistemological foundation, then God can obviously act naturally or supernaturally. Science and miracles can coexist in perfect harmony. But how exactly do we explain miracles in terms of scientific naturalism? Fortunately we don't have to. Supernatural events require no natural, material, or scientific explanation. They are outside the realm of scientific inquiry. The supernatural events that define Christianity such as the signs and wonders of God in the Old Testament, the miracles performed by Jesus, His resurrection from the dead and ascension into the heavens need no physical explanation from us. These signs and wonders speak for themselves. They were special cases where God, in order to demonstrate His power, caused events to transpire that were contrary to the natural order. If they had natural explanations, then they probably wouldn't have been that impressive.

These events described in the Bible were witnessed by many and recorded for others to be accepted or rejected by faith. Let's keep these precious truths out from under the jurisdiction of natural cause and effect and al-

---

[7] I've seen this same example used by others, including C. S. Lewis.

low the scientific method to do what it does best, explain the uniformity of nature in terms of the ordinary patterns of material behavior.

Healthy Spiritual Speculation

I'd venture to say that a lot of things today that are seen as "miracles" are probably just *coincidences* on the physical level. A coincidence is just a confluence of two or more events in a statistically improbably way. Sometimes it is good, like digging in the sand at the beach and finding your lost wedding ring from the year before. And sometimes it is bad, like having the only tree in your entire yard fall right on your car just as you are backing out of the driveway. But whether coincidences are good or bad depends entirely on how we perceive them. The actual chain of natural events that precedes a coincidence is merely the unfolding of the so-called undirected and unguided laws of nature. But we know that nature is only "undirected" and "unguided" from our finite perspective. God's complete sovereignty over creation gives direction to the course of natural history, even to ordinary coincidences resulting from secondary causes.

Think about some of the statistically improbable events in your own life that have occurred. These amazing coincidences may have seemed totally impossible, not because they required "divine intervention" to miraculously suspend the laws of nature, but because the odds were so stacked against them happening on their own. Somehow though, they managed to occur anyway, despite the odds. On the physical level, these things can't prove the existence of anything supernatural because they are just statistical coincidences, no matter how improbable. But the interesting thing about probability is that it is not a law like gravity. If I flip a coin ten times and get "heads" all ten times, no supernatural explanation is required.[8] Even though the "law" of probability says the odds of that happening are less than one in a thousand, nothing prevents me from getting ten heads or ten tails on the first ten flips.

One thing that Christians need to remember is that with God at the helm of creation, *improbable* is not the same as *impossible*. A practical application of this principle is that it does no good for Christians to try to prove something couldn't have happened naturally just by calculating its improbability. Probability might help us to investigate certain events on the material level, but it is not going to give us the whole story. Probability just tells us what we can expect to happen most of the time. Of all people, Christians shouldn't get too hung up over improbabilities. The Bible is full of them.

---

[8] Obviously the larger your sample size, the more coincidental the outcome. At some point, you would have to ask yourself if there is bias in your set up.

The doctrine of God's providential governance over the material world has very practical implications for how we approach our reconstruction of natural history. Consider the following argument carefully: if God wants to accomplish X via a material mechanism without suspending the normal patterns of material behavior, then X will happen regardless of the odds against it. Since X can be achieved through a natural process of cause and effect, X could leave behind physical clues to its occurrence. Such evidence could tell us a lot of useful information about *how* (but not necessarily *why*) X came about, even though all the details might not be clear. If science enables us to reconstruct these events and it becomes evident that X took place, then the odds against X happening become irrelevant at that point. *Why* such an unlikely event took place *despite the odds* may be an interesting philosophical tangent, but the responsibility of science at that point is to continue refining the theory of *how* X might have unfolded. Remember this principle for later.

When we see the natural order as the patterns of God's providence, even "ordinary" events like childbirth can take on profound meaning and purpose. We see God's hand behind everything, not just the so-called miraculous events in human history. We don't need to look for the likeness of Jesus on a piece of burned toast to remind ourselves that "God is there" because all of creation testifies to God's handiwork. The Scriptures make it clear that in some mysterious way God uses seemingly contingent events, like those governed by the laws of probability, to accomplish His purposes. The Bible says that God even works through the roll of the dice in a game of chance.[9] In this way, God subtly and quietly works out His divine providence without leaving any supernatural trace except for the spiritual meaning that results from the *context* of the event.

I am by no means the first to speculate on this. There are entire books about it if you are interested.[10] In fact, new discoveries in the field of complex dynamics and chaos theory show us how even the smallest change in a physical system can have large-scale consequences. Ever heard of the butterfly effect?[11] Again, this is not scientific proof of God's providence (of which such proof is both impossible and unnecessary), but it does allow us to engage in some healthy spiritual speculation of how God can interact with His creation without upsetting the fundamental coherence of nature. The important

---

[9] Proverbs 16:33

[10] Donald M. MacKay, *Science, Chance, and Providence* (Oxford; UP, 1978).
Donald M. MacKay, *The Open Mind and Other Essays* (Inter-Varsity Press, 1988).
John C. Polkinghorne, *Science and Providence* (Shambhala Publication, 1989).

[11] Complex systems, such as the many variables that contribute to the weather, are extremely sensitive to tiny changes in the initial conditions. So theoretically a tiny breeze caused by a butterfly's wing can lead to the formation of a giant weather system.

concept here is that trying to use science to prove God's presence is completely unnecessary. When God works *through* the laws of nature, no supernatural interference can be detected; and when God works *outside* of the laws of nature, He leaves no data for science to analyze.

Doesn't this whole idea of primary and secondary causes fit our everyday experience? We know God is active in the world, working His will throughout creation. But we also know that things mostly happen in accordance with the laws of nature, even the improbable things that everybody thinks are physically impossible. When the cancer that nobody thought would go away is finally in remission, we thank God for His mercy. However, we also know that on the physical level the tumor somehow responded to the treatment and we thank God for modern medicine as well. A miracle? Sure, but one that leaves no supernatural trace, violates no laws of nature, and yet still strengthens our faith and brings God glory. This is just another way of saying that God works in mysterious ways.

As I've said, these are theological ideas and not scientific theories. And theological claims can't be tested or proven in a laboratory. But I still see the results of these silly medical studies showing that sick people are healed in half the time when other people pray for them. Or maybe they get worse and die faster—I can't remember which. God just doesn't work like that. Spiritual promises are not formulas or laws that can be falsified or verified in scientific studies. We have to accept the fact that God's ways are more mysterious than that.

Things that appear to us as coincidences happen all the time and we sometimes call them miracles. But miracles that actually violate the laws of nature are probably few and far between. Other than the ones that were witnessed and recorded for us in the Scriptures, I would be pretty suspect of any such things today. I'm not saying they can't or don't happen, but they would definitely be the exception and not the rule. If people walked on water everyday and things were always changing into this or that, the world would not make much sense. There would be no physical certainty between natural cause and effect.

God is much more subtle than that. He quietly accomplishes His will using ordinary means and secondary causes, even mysteriously working through our free choices.[12] I believe that His "hand" of providence is holding me down in my chair right now. Newton's law of gravity can help me quantify it, but in my mind, Newton's law of gravity can never replace providence. These claims are not in competition with one another. And neither can any

---

[12] The story of Joseph in Genesis is my favorite example of how God works through the free choices of individuals to accomplish His will on earth.

other material mechanism supersede a transcendent purpose because they are two different answers to two different questions about reality.

The Laboratory versus the Courtroom

Many people really only care about science in so far as it makes their lives easier or provides some sort of entertainment. And when it comes to entertainment, it is interesting to note the rising popularity of television shows on the topic of forensic science. Almost any night of the week, pure science is being pumped into our living rooms and it is keeping us on the edge of our seats. Does anybody else find it ironic that we can be so captivated by the science of "stuff" like bones, bullets, hair, fibers, and DNA?

The reason we love it so much is because these programs demonstrate how powerful science can be. We can hand out life or death sentences for crimes that nobody witnessed, where no body was found, and that may have taken place long ago. For people who may not have even considered anything scientific since high school, that is pretty incredible.

Everybody loves it when a detective takes one of those really old "cold" cases that nobody could solve for years and cracks it wide open. How can science take a bunch of seemingly unrelated facts and tell us a story in which each piece of information is like a main character acting out the events that connect a crime with a perpetrator? Because of recent advances in modern technology, the same old evidence can now reveal mountains of new information that was previously unavailable.

The revolution in forensic science is driven by many of the same principles that are revolutionizing the ongoing scientific reconstruction of natural history. Amazingly, every single piece of "stuff" has a story to tell, if we only know what to look and listen for. Every object carries the clues of its natural history with it. In the laboratory, things can tell us what they're made of, when they were made, where they came from, where they've been and what happened to them along the way. A sample of DNA from even the smallest bit of hair or tissue can identify a person, a bullet dug out of a wall can identify the firearm that shot it out, the type of paper and ink of a document can identify where it was printed, the wear and position of letters on a typed page can identify the typewriter that typed it, pieces of a shredded document can identify the shredder, and the list goes on. When the natural history of every piece of evidence is revealed, it should tell a consistent story that either condemns or exonerates the accused.

In the courtroom, there is a slightly different drama playing out. The objective here is greater than simply the guilt or the innocence of the accused. The objective here is justice. The transcendent concept of justice is beyond the limits of science. Justice can't be determined in a laboratory by following procedures. That's what judges and juries are for. Here we see both the importance, and the limits of the scientific method.

In the laboratory, only a coherent natural explanation of the physical evidence is acceptable. The prosecution lays out the facts of the case in chronological order and builds some kind of timeline showing how each piece of evidence fits into a consistent pattern. This establishes a logical chain of cause and effect that leads directly back to the accused. The defense must deal with the facts of the case by attacking the coherence of this logical chain of events, pointing out weakness and offering alternate explanations.

If the prosecution were to invoke any kind of supernatural occurrence to cover gaps in the timeline just to make the charges stick, the defense would tear it to shreds and the case would be over in a heartbeat. Can you imagine what a circus that would be? "Well, Your Honor, I know the accused was out of town on business the night of the murder, but we must consider the possibility that he teleported himself back home where he committed this murder before teleporting himself back out of town." Not even deeply held religious convictions are allowed. If the defense were to argue from Scripture that the accused was justified in burying the victim under a pile of rocks because he uttered a blasphemy[13] against God Almighty, then the jury would never even be allowed to consider it in the guilt or innocence phase.

On the other hand, science can only go so far before giving way to metaphysical arguments. Once the facts of the case are explained, the jury must consider motives, mitigating factors, premeditation, mental status, and a host of other value judgments that start to move them from the domain of science to the domain of theology, tradition, and philosophy. In the end, the jury must act on the physical evidence, but in doing so they must ask, "What does this mean?" They must make value judgments.

If the defense decides in the closing arguments of a murder trial to argue that competition exists among members of the same species for limited resources and that the killing of the victim was simply the result of the process of natural selection, the judge would probably not even allow this line of reasoning. Even if it could be proven, beyond a reasonable doubt, that the removal of the victim's genes from the human gene pool by the defendant was good for the species, no jury outside of Los Angeles would even consider it. The truth of it has no bearing on the requirements of justice. The responsibility of the defendant to live in peace with his neighbors transcends his biological development. We are greater than the sum of our parts and experiences, and we are commanded to act like it. In fact, Jesus instructs his followers to

---

[13] Leviticus 24:16

feed the poor, help the weak, comfort the sick, shelter the homeless, and visit the prisoners. This is not exactly the "survival of the fittest."[14]

Science deals with the "raw" data, the "purposeless" laws, and "meaningless" facts. When it attempts to provide meaning or make value judgments of an immaterial nature, it has probably crossed a line somewhere. In practice, there are arguably some areas of overlap. Things like mental illnesses, chemical imbalances, and environmental factors are all controversial because they blur the line between the physical and the metaphysical.

There are important lessons here. In the laboratory, scientific naturalism is absolutely essential to explain the facts of the case. The lab technicians don't care about the outcome of the case, their objective is to follow the procedures and adhere to the standards of laboratory science. Their personal, philosophical, and religious commitments are about as relevant as what they ate for breakfast. Only cold hard science is allowed to explain the physical cause and effect relationships that connect the victim with the accused. Supernatural explanations are never even considered.

However, in the courtroom, the jury must determine what this evidence means. They must look beyond material cause and effect to make value judgments based on ethics, morality, extenuating, and mitigating circumstances. Their personal beliefs and value systems will undoubtedly guide their interpretations of the facts. That is why the jury selection process is so important.[15] Each side wants folks on the jury who will not be predisposed against their interpretation of the facts. Here we see both the usefulness and limitations of the two ways of knowing.

Looking Ahead

Obviously the application of scientific naturalism, theology, and philosophy to the creation/evolution controversy has its own unique set of challenges. The distinction between material mechanism and transcendent meaning that was so clear in my overly simplified and carefully chosen examples starts to get pretty complicated when we look at the biblical creation narrative in Genesis. The Bible seems to be giving us both the mechanism and the meaning woven together into six distinct 24-hour days. The primary causes and the secondary causes almost seem to overlap. Other passages sprinkled throughout

---

[14] In the spirit of fairness, some evolutionary biologists claim that altruistic behavior can be beneficial to the tribe and is therefore able to be passed down by natural selection. Some of the arguments are compelling, but regardless of whether or not this is true, human nature is what it is. Jesus clearly taught how we are to treat others, period.

[15] And that is also why there is not a "laboratory selection" process, because the methodology of laboratory science is not subject to personal opinion or philosophical bias.

the Scriptures also appear to give us literal scientific information about the cosmos. Trying to untangle this web is no easy task and five different Christians will probably give you six and a half different answers. But I hope this section of the book has at least caused you to think epistemologically. That is, to consider what we know and how we know it, so that you can draw your own conclusions when sifting through the mountains of conflicting arguments out there.

In the following chapters, we will apply some of these basic ideas to the question of our origins and see where that takes us. We will look at two different sources of knowledge: the Bible (special revelation) and nature (natural revelation), in that order. We open the book of *God's Word* and the book of *God's works*, taking each one seriously in an honest attempt to understand what they are telling us. With the lessons of history as our guide, we will try to piece together a consistent understanding that doesn't put one at odds with the other.

PART II

WHAT CAN THE BIBLE TELL US ABOUT NATURE?

CHAPTER THREE

THE CONTEXT OF CREATION

The point of this chapter is not really to argue the authority or infallibility of the Scriptures. You can find many resources on that topic and nothing I could say would add anything of value to those. The Bible is a self-validating authority that can't be justified by appealing to anything outside of itself, including science, but on what exactly is it an authority? Are Christians at liberty to ask of the Scriptures any type of question that they wish? Will the Bible help me do my taxes or instruct me on how to rebuild a carburetor? Of course not. So right away, we can acknowledge that perhaps not every question has a straightforward "biblical" answer.

We can also all agree that for Christians, the Bible is the one and only guide to the supernatural. Without it, our attempts at connecting with the spiritual realm would look no different than witchcraft or black magic. The Bible is our primary source of knowledge about God, the condition of man, and the person and work of Jesus Christ. Ideas about right and wrong are not always so easy to determine, but we can at least agree that these things are transcendent realities and as Christians we look to the Bible, rather than to nature, for clues about their understanding. We believe the Bible to be nothing less than divine revelation. But what exactly is divine revelation and how do we use it properly?

How Did We Get the Bible?

The Bible claims to be the inspired Word of God.[1] So how did we end up with it? Does God have a publishing contract? Not quite. The Bible was written by the hands of men under divine *inspiration*. But what does that actually mean? Does it mean that every word of the original text was specifically chosen by God? Some say yes, but that would seem more like divine *dictation*. Or does inspiration mean that while the general ideas and themes were impressed upon the biblical writers, they actually wrote down *in their own words* what God had laid upon them?

---

[1] II Timothy 3:16

On the surface, dictation would seem to reduce these great men of God to human fax machines with God sending the data. Or it puts them in some kind of New Age trance, wielding a writing instrument like a divine Ouija Board. The fact that each writer has a distinct literary style seems to indicate that there was some artistic license in terms of the actual words chosen. But then again, some important Christian doctrines seem to hinge on the exact interpretation of a few key words of the Scriptures. And you can't really trust something this important to the writing abilities of sheep herders, goat farmers, tent makers, and fishermen can you?

The answer to this mystery probably lies somewhere in the doctrine of God's providence. Perhaps God used ordinary means to reveal everything we needed to know about the Kingdom of Heaven without compromising the free will of individual human authors. In some mysterious way, the words belong to both God and the original authors. And so the divine purposes of God must somehow be expressed through the human purposes of the author. This makes the *immediate* intent of the biblical writers our most accessible link to the *ultimate* intent of the Holy Spirit—making God's Word both *timely* (relevant to the original audience), and *timeless* (relevant to all generations).

In order for God to communicate His divine truth to us, it had to first be put into words for us to understand. It had to be *accommodated* to an existing human language. This sounds easy enough, but the fact that a particular language among the many available languages was chosen is quite remarkable. So already we see God using an existing cultural form (timely) to express His eternal truth to every generation of every culture (timeless). This might seem like trivial information, but consider this: if we didn't have the Bible translated into our language, we wouldn't know what it says unless we first learned Hebrew and Greek. And not just modern Hebrew and Greek, but the ancient versions—complete with all their ancient literary and cultural nuances. That would, in turn, require us to know quite a bit about the ancient Hebrew and Greek beliefs, customs, and traditions. Their modern equivalents would be completely irrelevant to our understanding of these ancient texts.

I personally don't know the first thing about ancient Hebrew or Greek, but I know a little something about English and it's not the easiest medium for effectively communicating timeless truth across many diverse cultures. Consider the regional variations. In Great Britain, a cigarette is a *fag*, an elevator is a *lift*, the trunk of a car is the *bonnet*, a bathroom is a *water closet,* and getting *pissed* has nothing to do with your temper. And even within Great Britain, you have a Scottish dialect, an English dialect, an Irish dialect, and a Welsh dialect. If you also consider that Australia, Canada, New Zealand, South Africa, the Caribbean and others all have their own unique vocabularies, figures of speech, accents and dialects then it becomes even more ridiculous.

Add to that the differences in any particular dialect through time. Take American English for instance. Certain phrases and figures of speech commonly used today would not make any sense a century ago. If I told somebody from the past that we're going to the beach to *catch some rays*, they might think we are marine biologists studying stingrays in the surf. If I say that something won't happen until *hell freezes over* or until *pigs fly*, it doesn't mean that these events are inevitable. But somebody not familiar with these figures of speech might incorrectly assume we've found a way to extinguish the fires of hell and pigs have evolved into birds.

So is the Hebrew language immune from these complications? Is the Greek? I don't think so. I'm sure that dialect, metaphor, figure of speech, slang, and the cultural context of these original languages all pose similar challenges to interpretation. These are the kinds of difficult issues that Bible translators have to deal with in order to give us something that makes sense in our time and in our language. Understanding this should give us a deep appreciation for those who toil under such difficult circumstances to bring us these timeless truths. It should also give us a measure of humility and caution us about being so certain that a specific interpretation is accurate.

So is the Bible infallible? Absolutely! What about inerrant? You bet! Easy to understand and apply? Not quite. All I'm saying is that there needs to be more humility across the board when Christians open these sacred books of the Bible, especially when it comes to the difficult details of creation.

How Do We Use the Bible?

One of the most important things to remember is that divine revelation should have made sense to the original audience to which it was given. No matter how strange it might seem to us today, our interpretation must be something that they could have understood given their perspective. Of course this doesn't mean that they always did. In fact, we see many times where God's people just didn't get it. But the point is that the original meaning of the text wouldn't have been obscured by some futuristic context that made no sense to ancient people. We shouldn't sacrifice the *timeliness* of the Bible for the sake of its *timelessness*. Each generation that received divine revelation would have had everything they needed to properly understand the message, provided they had "eyes to see" and "ears to hear."

This idea might seem like a no-brainer, but how many times have you heard a wild interpretation of Scripture that would only make sense to a modern audience? Some popular interpretations of the book of Revelation include modern armies, tanks, helicopters, current events, 20th century leaders and nuclear weapons. You've probably seen some of these things for yourself. Now for some, this can be fun and provide hours of entertainment, but eventually you have to ask yourself how a first century audience would have ever understood that! In general, if an interpretation of Scripture requires specific knowledge that would have been unknown or unavailable to the original audience, then you probably need to start over.

Along those lines, I've come across at least two different interpretations of Genesis 1 that try to fit modern cosmology into the literal six-day framework. In other words, the Big Bang happened just as science says it did, but to avoid the "embarrassment" of the six-day biblical account we bend and twist the Scriptures to show that these ideas were right there in Genesis all along. This kind of stuff makes for fascinating reading, but these interpretations require knowledge of advanced concepts like the *cosmic background radiation*[2] and *gravitational time dilation*[3] in order to fully understand the meaning of the text.

Now I'm not going to get into a detailed discussion on whether or not these specific interpretations are theologically or scientifically sound, but the thing that immediately strikes me as problematic is that the ancient Hebrews wouldn't have had any way to understand these modern scientific concepts. So in order to accept these interpretations you'd almost have to conclude that after Moses wrote down the text, the "true" meaning of Genesis 1 remained a secret for thousands of years, until 20th century astrophysics finally gave us the tools to properly decipher it. And if that is really the case, then these passages of Scripture would have been misleading to both the original audience and the many generations of believers thereafter who just read them at face value, each in the context of their own day.

I think that this super-scientific approach to interpreting Scripture misses the point entirely. When we force something from the Bible to fit into a mod-

---

[2] The cosmic background radiation is the microwave "static" that pervades all of empty space. It is believed to be a relic of the Big Bang that originally included visible light, but the subsequent expansion of the universe stretched out the light waves into microwaves. See Gerald L. Schroeder, *The Science of God: The Convergence of Scientific and Biblical Wisdom* (New York, NY; Broadway Book, 1997).

[3] Gravitational time dilation is a consequence of Einstein's theory of relativity and says that if you take two identical clocks and place them in two different gravitational fields, they will tick at two different rates. A few seconds on one clock could represent millions of years on the other depending on the relative strength of the gravitational fields. Time Dilation is not just a theory, but a fact that has been experimentally verified. Some interpretations of the six-days of creation require the stretching of time in different reference frames in order for the universe to fully evolve according to modern cosmology in six days. Apparently Moses forgot to mention this so the key to understanding the first book of the Bible would not be discovered for another 3,500 years! See D. Russell Humphreys, *Starlight and Time: Solving the Puzzle of Distant Starlight in a Young Universe* (Colorado Springs, CO; Master Books, 1994).

ern scientific context, we exchange the timeless truths of the Christian religion for the ever-changing theories of natural science. We must also consider the unintended consequences of our bending and stretching the Scriptures to accommodate a specific scientific theory. This is a dangerous and slippery slope indeed. If we can read new ideas into Genesis to fit a modern scientific theory, then what would stop us from reading new ideas into any other part of Scripture to accommodate some kooky social theory? I think we all know what kind of abuses that could lead to. And what happens when scientific paradigms shift, as they inevitably do? Will Christians then be left scrambling to adjust their theology to accommodate another scientific theory? This is absurd.

We also should avoid sacrificing the *timelessness* of the Bible for the sake of its *timeliness*. In other words, divine revelation must have also been written in such a way that its meaning can be properly understood by all future audiences. Again, this may seem like a no-brainer, but how many times have you seen Christians dismiss certain passages of Scripture because there doesn't seem to be any relevant modern application? While it's true that many things like the Jewish system of bloody sacrifices and dietary laws are no longer relevant to the church in terms of procedure, they can still teach us volumes about the character of God if we know what to look for. Usually this involves trying to understand the original context in which the text was received, separating the timeless principles from the cultural forms used to convey those principles, and contextualizing these precious truths for our time. I'm not implying that it is always easy, but at least it allows the entirety of God's word to speak to us in our time.

Context is Everything

On the surface, the Bible is a strange piece of literature. If you were to completely remove the cultural, historical, and geopolitical contexts, it's actually a pretty offensive document by modern Western standards. A casual reader might conclude that slavery, polygamy, animal sacrifice, genocide, ethnic cleansing, concubines, witchcraft, discrimination against disabled persons, male chauvinism, and a host of other antisocial and dysfunctional behaviors are all a normal part of the Judeo-Christian life.

The concept of slavery for starters is absolutely contrary to Christian principles, yet the Bible appears to be more concerned with regulating slavery than abolishing it. The Bible clearly upholds the one man/one woman concept of monogamy,[4] so what's the deal with these Old Testament patriarchs

---

[4] I Timothy 3:2

having harems of wives and concubines and not even getting busted for it? Love your enemies you say? How about God's command for Israel to totally wipe the Canaanites and Amalekites from the face of the earth?[5] And if you think only the Old Testament is offensive, what about the New Testament command for women not to speak in church?[6] Or the prohibition against women having short hair or being without a hat in public?[7] And slaves being ordered to submit to their masters and take their beatings patiently?[8]

When these difficult verses are ripped from their original contexts and transferred to modern situations, the results are never good. In fact, many atheists—woefully ignorant of ancient Near-Eastern culture—think they are making clever arguments against Christianity when they attempt to point out how obviously "out of sorts" the Bible is with our own society.[9] These pathetic attacks never include any scholarly attempt to understand the complex social milieu of the ancient Near East. They completely ignore the *timeliness* of the ancient Scriptures and instead try to directly relate them to modern social situations without making any adjustments in their expectations. But before we climb up on our high horse, we should first recognize that many Christians are guilty of doing this very thing when it comes to understanding creation. We completely ignore the unique set of circumstances surrounding Genesis and instead try to relate the text directly to a modern understanding of the cosmos. And just like our atheist counterparts, we make no adjustments in our expectations of the text.

So in what context was the book of Genesis received? The first five books of the Bible are believed to have been written by Moses sometime after he left Egypt and before his death. Assuming that he started it soon after leaving Egypt, it's quite possible that the creation accounts of Genesis 1 and 2 were already complete around the time of Mt. Sinai. As with any book of the Bible, the date and author are disputed, but I'm not going to open that can of worms here. Regardless of the human author and the exact date, the ancient Near-Eastern cosmological context doesn't change much until the rise of Hellenistic astronomy over a thousand years later; and that's all that really concerns us here.

About 500 years passed between God's covenant with Abraham and Moses' hike on Mt. Sinai. During this time, the Hebrew people didn't have any of the Old Testament Scriptures as we know them. Instead, God spoke to

---

[5] Deuteronomy 20:16-18
[6] I Corinthians 14:34, 35
[7] I Corinthians 11:5-10
[8] I Peter 2:18-20
[9] For a perfect example, see Sam Harris, *Letter to a Christian Nation* (New York, NY; Alfred A. Knopf, 2006).

the people through the Prophets and those words of wisdom were probably passed down from one generation to the next by a combination of oral tradition and other writings that are sometimes referred to in the Bible.

Before Joseph moved his family to Egypt, the Hebrews lived among the various gentile cultures in Mesopotamia. We know from the histories of both the Mesopotamian and Egyptian cultures that they each had elaborate creation myths and detailed *cosmogonies*[10] explaining how the universe was built and how it worked. These were based on naked-eye observations of the physical universe and animated by polytheistic pagan religions. Likewise, the Hebrews probably also pondered their place in the cosmos and how the world around them was constructed. All ancient cultures did this, even the Native American people. And since these ancient ideas were usually based on how the universe appeared from the vantage point of earth, all primitive cultures had very similar ideas about how the universe was constructed and how it worked.

So it's probably safe to assume that after living in the ancient Near East for hundreds of years surrounded by these pagan cultures without any written Scriptures, the Hebrew people adopted some of the common knowledge found in that region. In fact, we know from the Bible how easily the Jews were influenced by the surrounding nations because throughout the Old Testament we see God constantly reminding His people to remain separate and distinct from the gentile people that dwelt among them. The Jews were not to worship their pagan gods, marry their heathen women, or adopt their customs. They obviously had a habitual problem with this or God, speaking through the prophets, wouldn't have spent so much time reminding them of it. The laws of separation that forbid them to mix different seed in the field or weave different threads in their garments were to be a constant reminder not to integrate with the pagan cultures around them.

In light of these various pagan creation mythologies and their obvious influence on the Hebrews who were still using oral tradition to pass information from generation to generation, it should not come as any surprise to us that the first thing God inspires Moses to write is the "official" creation story from the Creator Himself. But before we look at what the Bible can tell us about the beginning of time, it might be helpful to know what those Mesopotamian and Egyptian creation ideas were. The reason for this should be obvious; this is what Moses was up against. Those pagan mythologies provided the immediate contemporary backdrop against which Moses recorded the Hebrew account. And just like the Bible translators who have to first learn

---

[10] A *Cosmogony* is a non-scientific account of how the universe came to be and how it operates; as opposed to a *cosmology*, which is a scientific account.

the ancient languages, we must familiarize ourselves with the ancient creation myths in order to understand the specific issues that Moses might have addressed in Genesis.

One very insightful biblical commentator and ancient Near-Eastern scholar, Dr. John H. Walton, puts it like this:

> We live in a world far different from the world of the Old Testament. We must recognize the elements that distinguish these two worlds and make appropriate adjustments to our expectations. In our world, we believe reality is described most accurately in scientific terms. Mythology in the ancient world played the role that science plays in our modern world—it contained the explanation of how the world came into being and how it worked.[11]

So how do we familiarize ourselves with ancient ways of thinking and train our modern minds to properly understand Genesis? Dr. Walton continues:

> The conclusion, then, is that we can often identify the questions the text addresses by familiarizing ourselves with ancient literature rather than by letting *our* culture dictate what questions the text addresses or how it answers questions. Once we stop pressuring the text to address our issues, we may find that it is easier to identify the text's issues.[12]

As you can imagine, mythology is not a medium of communication that most modern people are familiar with, but if we make an effort to study these ancient Near-Eastern stories, they can offer us a very useful window into the world of the ancient Israelites. Quite frankly, they are the primary cultural context in which the Genesis creation account must be understood.

The Ancient Near-Eastern Universe

The Mesopotamian creation myth is called the Enuma Elish. You can read the whole thing in a library or on the internet so I'm just going to hit the highlights.[13] The story was shared by the Babylonian, Assyrian, Akkadian, Chaldean, and Sumerian people, who basically occupied the same lands as the

---

[11] John H. Walton, *The NIV Application Commentary: Genesis* (Grand Rapids, MI; Zondervan, 2001), pg. 83.

[12] Ibid, pg. 84.

[13] Alexander Heidel, *The Babylonian Genesis 2nd ed.* (Chicago, IL; University of Chicago Press, 1951).

Hebrew people before they were taken into Egypt. In fact, many of the Hebrews, including the patriarchs, would have come from these lands, bringing much of their cultural baggage with them.[14]

According to these ancient mythologies, the heavens and the earth were created by a multitude of gods. Acting more like a dysfunctional family than a host of noble deities, they began creation by mixing together the fresh and salt waters that would make up the great ocean. Interestingly, these gods did not create the world out of nothing, but rather they just rearranged existing matter into the heavens and the earth. These gods then seem to quarrel among themselves for quite a while, raising armies out of the primordial abyss to fight against one another to the death.

Eventually, the leader of the gods from the winning side, Marduk, carves up the bodies of the gods from the losing side and starts to build the earth and sky out the carcasses. The leader of the defeated gods, Tiamat, is split in half and her body is used by Marduk to divide the waters above the earth from the waters below. So half of her body becomes land, resting on the watery abyss, and the other half of her body becomes the top of the sky, which is supported by great stands, or pillars, memorializing some of the other gods.

The moon and stars are then created from the corpses of the defeated gods and they hang from underneath the waters above. Gates were opened in the sides of the sky between the ribs of Tiamat's carcass so that the sun and moon could come and go each day. There is also an interesting description of the ancient calendar of 7-day weeks, 30-day months, and 12-month years being established by these heavenly bodies. The lower half of Tiamat's carcass, which became the land, was shaped into mountains, valleys, and plains. Her eyes were plucked out from her skull and out came the Tigris and Euphrates rivers. Primitive man was then formed from the blood of one of the executed gods and the Babylonian Empire was created so man can serve the gods. There is a little more to it than that, but that's the gist of it.

As fascinating as all of that is, the generation of Hebrews that received the Bible directly from Moses might have been more familiar with the Egyptian creation myths. Moses, being educated in the best Egyptian schools, would have certainly been very familiar with all the wisdom of Egypt, including their many creation accounts.[15]

According to the Egyptians, before there was even heaven and earth, there was a primordial sea representing the state of chaos and disorder. *Chaos* was associated with the destructive forces of nature that primitive man constantly lived in fear of. We don't typically look at nature as a delicate balance between order

---

[14] Joshua 24:3
[15] Exodus 1:15-19; 2:1-10; Acts 7:22

and chaos, but this is very common in more primitive societies. Except for an occasional natural disaster in some other part of the world that makes the evening
news, modern Westerners live relatively comfortable lives, sheltered by our technology from the dangers of creation.[16] But ancient man was under constant threat
of famine, pestilence, flood, drought, storms, wild animals, and earthquakes.
Creation by the gods is therefore achieved, not necessarily by building things, but
by imposing *order* on the cosmos—effectively restraining the forces of nature
for the sake of human civilization.

From this initial dark and watery state, ruled by the god Nun, a fertile mound
of dirt emerged along with the god Atum, who held the creative power to bring
order out of the primordial chaos. For any culture living along the Nile, this image of creation emerging from a watery state would have been associated with
the annual flooding and receding of the river. The universe was thought to have
been born from similar floodwaters as part of an ongoing cycle of creation and
re-creation by the gods.

There were several regional versions of the Egyptian creation myth and
each one had a different name and made reference to different local gods, but
the common theme is that more gods were then created by Atum to help order
emerge from the chaos by a series of separations. The land and the sea were separated, as was the sky from the earth. The goddess Nut, representing the heavens, is often portrayed as a naked female with stars on her body arched over the
god Geb, representing the land, with her hands and feet resting on each of the
four corners of the earth. Every morning, Nut gives birth to the sun god, Re,

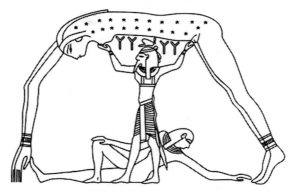

Figure 1: A typical depiction of the sky god Shu holding up (or separating) the heavens (the goddess Nut) from the earth (the god Geb).[17]

---

[16] Hurricane Katrina in August of 2005 reminded us all about the chaos and destruction of the
"primeval" waters.

[17] Barry L. Bandstra, "Reading the Old Testament: An Introduction to the Bible" Chapter 1,
http://www.hope.edu/bandstra/RTOT/CH1/CH1_1A3A.htm

who passes underneath her body before she swallows him each evening.

The details of the Egyptian creation myths are slightly different than the details of the Mesopotamian myth but there are some obvious similarities, like the idea of the heavens and the earth emerging from a watery abyss and the physical bodies of the gods making up the land and the sky. Both mythologies also viewed the sky as a solid structure separating the waters above the heavens from the waters below, and they both shared the idea that the sun entered and exited the sky through openings in the bodies of the gods.

After hearing these ancient Near-Eastern creation stories for the first time, many readers may be wondering how anyone could possibly believe such outrageous things. Did the Tigris and Euphrates rivers really flow out of the defeated Tiamat's eye sockets? Was the sky really a goddess arched over the land with her hands and feet resting on the four corners of the earth? Were hallucinogenic drugs really that common in the ancient Near East? But questions like these only reveal our modern bias, and they demonstrate just how radically different our modern Western worldview is from the ancient Near-Eastern worldview. Dr. Walton again illustrates this point when he writes the following:

> When *we* ask the question "How does the cosmos work?" we seek an answer that discusses physical laws and structures, matter, and its properties. In our worldview, function is a consequence of structure, and a discussion of creation therefore must, *of course*, direct itself to the making of things. In contrast, when an Israelite asked "How does the cosmos work?" he or she was on a different wavelength, because in the ancient worldview, *function is a consequence of purpose*.[18]

If we know what to look for, these unbelievable mythologies can show us how ancient cultures related the known functional elements of the cosmos to their divine purposes. Apparently, ancient man was not the least bit concerned with the naturalistic demands of some future culture such as ours.

Why should we concern ourselves with this? Because there is little reason to believe that the Hebrew people—who were as much a part of the ancient Near-Eastern social landscape as any of these other cultures—would have described the cosmos in any other terms. So again, the creation mythologies of Israel's geographic neighbors can provide us valuable insight into the Hebrew mindset, helping us place the Genesis creation account in a timely context.

---

[18] John H. Walton, *The NIV Application Commentary: Genesis* (Grand Rapids, MI; Zondervan, 2001), pg. 85.

These pagan creation mythologies gave rise to a popular conception of the physical universe that wasn't really challenged in that region until the rise of Greek astronomy thousands of years later. The earth was seen as a flat disk or a rectangular table top floating in a vast ocean. Heaven was seen as a solid dome, or vault, which arched over the earth and supported another body of water above the sky. The vault of heaven was supported by pillars—thought to be great mountains—whose foundations were laid in the great waters surrounding the earth. The waters above the sky were continuous with the waters around and under the earth. The firmament, being a solid structure, had doors on the east and west sides through which the sun, moon, and stars would enter and exit each day. The rains were caused by tiny windows in the firmament that let down some of the waters from above the sky.

You might think this model of the physical universe is ridiculous, but understood from within the context of the non-scientific ancient world, it was actually fairly consistent with general observation. The earth does look flat from our perspective. On a clear day if you walk outside, the sky sometimes looks like a far away ocean. And the fact that water comes from the sky when it rains is strong "evidence" in support of this observation. When looking at the motion of the sun, moon, and stars across the sky, they appear to move in a semi-circle from one end of the horizon to the other while the earth seems fixed and immovable. So it would have been perfectly natural to assume that the shape of heaven was that of a dome or a vault with the celestial bodies "moving" across it. Have you ever been to a planetarium? The easiest way to re-create the night sky is to build a dome and project lights on the underside. And since the heavens and the waters above seem to be unaffected by the earth's gravity, they must be supported by a rigid, yet invisible structure (the firmament) that also regulates the passage of wind, rain, and celestial objects through the sky.

By non-scientific standards, the ancient cosmogony was a pretty decent description of the physical universe, even though we now know that it was horribly inaccurate by today's standards. Nevertheless, it is against this backdrop that God gave Moses the straight skinny on creation. This was God's opportunity to correct almost 500 years of misinformation and set the creation record straight. Are you prepared for what comes next?

A Beautiful Letdown

The biblical account starts out with the most foundational of all creation claims: that the Triune God created the universe. Set against these other creation accounts, this would have been a very serious statement. According to the Enuma Elish, the world was borne out of a conflict between rival gods, and was therefore subject to the whims and wants of temperamental deities with conflicting interests. In this chaotic and unpredictable world, humanity was sure to get caught in the crossfire. Old gods are being killed and new gods are being born. Each new divine administration brings more uncertainty and unpredictability to nature. In sharp contrast, the Bible shows us a coherent world created in unity by the three persons of the Trinity working together in total harmony. Order, purpose and uniformity are built into the universe from the very beginning.

Right out of the gate Moses throws down the cosmic gauntlet by challenging the gross polytheism of the pagan mythologies. But that was merely the first sentence of the Hebrew creation story. What happens next can almost seem like a letdown if we have any misguided expectations of a modern scientific treatise on the construction of the observable universe. Rather than seize the opportunity to overturn the commonly held view of the universe which was riddled with theological and cosmological error, God seems to hijack the popular cosmogony and use it as a vehicle to set the *theological* record straight, leaving the *cosmological* record intact.

In other words, God gives Moses a new *theology of creation*, not a new *creation science*. In fact, while the Scriptures don't demand that we still believe in the ancient Near-Eastern cosmos, this model of the universe is never disputed by the Scriptures. These strange conceptions of the "heavens and the earth" may have seemed correct by ancient non-scientific standards of simple observation, but the whole idea that the earth was a flat disk or great big table sitting over a watery abyss and lying under a solid firmament is obviously incorrect by modern scientific standards. And the water that surrounds the solid sky and all heavenly bodies clearly does not exist. Why wouldn't God take this opportunity to tell us about the spherical earth, the solar system, the vastness of space and the proper size and position of the stars? And why are there no planets, nebulas or galaxies? Why didn't God settle these issues when He had the chance? Why did God just recycle the erroneous structural elements of the ancient Near-Eastern cosmos and intentionally keep mankind ignorant of the scientific truth for thousands of years?

These are all legitimate questions, or at least they should be. Try asking them at your church and folks probably won't be too happy with you. In fact, you may end up at the top of somebody's prayer list. But I don't know what else to tell you. You can pull out your Bible and follow along if you want. The second sentence of the biblical account starts off with the same formless watery abyss as the pagan versions.[19] Now think about that for a minute. Genesis 1:2 makes a reference to "the surface of the deep," but no explanation is

---

[19] Genesis 1:2

provided for this primordial structure. What surface of what deep? There are no footnotes or references, and there is no preface to explain what this watery abyss is all about. Why is that? Quite simply—it wasn't necessary! The Hebrews had just spent over 200 years in Egypt. They would have all known exactly where Moses was going with this. And just like the Egyptian accounts, God brings order out of the primordial chaos by a series of separations. He separates light from darkness and calls them day and night.[20] He separates dry land from water and calls them land and sea.[21]

This all seems like a strong indication that the construction of the ancient Near-Eastern universe was also the framework for Hebrew creation. But in my mind, the biblical firmament is the smoking gun. What am I talking about? Here we have a very literal description of something that was thought to physically exist in the ancient world. The dome that was above the ancient sky was a material entity, not figurative or metaphorical of something else.[22] It was a solid structure capable of supporting the ocean above it. But rather than correct this common misunderstanding that the sky holds back an ocean of water above the heavens from which it rains through tiny windows, God creates this strange thing called a "firmament" to separate the waters above the earth from the waters below the earth, thus dividing the "waters from the waters."[23] Here we clearly see God intentionally repeating the same structural inaccuracies of the pagan cosmogonies.

Now if the purpose of the biblical account of creation was to give the Hebrew people an accurate description of the physical universe, then this whole "waters above the sky-dome" thing should have been left out. In fact, many of the physical errors of the ancient cosmos should have been omitted; and there are several elements of the modern universe that could have easily been included—if that were indeed God's intent. So here are our options: either (1) God is poor communicator who knows less about the universe than we do, or (2) giving 21st century Christians accurate scientific information about the material universe is not the true purpose of Genesis. Which seems more probable? My vote is with (2)!

This might seem like an arrogant thing to say, but take a step back and think about it for a minute. If there were ever a time for God to reveal a correct cosmology to His chosen people, this was it. Here we have what seems like the perfect opportunity to put the scientific smack-down on the ancient underwater "planetarium," stick the sun, moon, and stars in deep space where

---

[20] Genesis 1:4, 5
[21] Genesis 1:9,10
[22] Paul Seely, "The Firmament and the Water Above," *Westminster Theological Journal 53* (Philadelphia, PA; 1991), pp. 227-240.
[23] Genesis 1:6-8 (KJV)

they belong and give the earth an atmosphere. Some planets would also be nice. But instead of dismissing the ancient firmament, the Bible says that God calls this structure a "sky" and places the heavenly bodies *in the sky and under the waters* to be signs for seasons, days, and years.[24] It would be thousands of years before the church, under pressure from Hellenistic astronomy, would completely give up this model of the physical universe.

Many modern commentators try to downplay the fact that the biblical firmament was a solid structure analogous to the ancient sky dome, but only because we all now "know" that a solid firmament does not exist. Hindsight is always 20/20. Since all of modern Christendom has moved beyond the ancient firmament, we now conveniently reinterpret these verses to fit modern astronomy. But this wasn't always the interpretive approach to the biblical firmament. Before the rise of modern science, many Christians believed in a solid sky and the waters above it simply because that *is* the most straightforward interpretation of Scripture. The early church fathers, who were obviously not influenced by modern astronomy, interpreted the Bible without any scientific bias. Origen (A.D. 185–254), one of the early church fathers, declared that the firmament was "without doubt firm and solid."[25] St. Ambrose (A.D. 340–397) confirmed that "the specific solidity of this exterior firmament is meant"[26] and St. Augustine (A.D. 354–430) said that the firmament "constitutes an impassable boundary between the waters above and the waters below."[27]

This was not just a problem for the early church either. Martin Luther (1483–1546), during the time of the Reformation said the following:

> Scripture simply says that the moon, the sun, and the stars were placed in the firmament of heaven, below and above which heaven are the waters…It is likely that the stars are fastened to the firmament like globes of fire, to shed light at night…We Christians must be different from the philosophers in the way we think about the causes of things. And if some are beyond our comprehension like those before us concerning the waters above the heavens, we must believe them rather than wickedly deny them or presumptuously interpret them in conformity; with our understanding.[28]

So according to Martin Luther, we are all guilty of "wickedly denying" the

---

[24] Genesis 1:8, 14-18

[25] First Homily on Genesis, FC 71

[26] Hexameron, FC 42.60

[27] The Literal Meaning of Genesis, ACW 41.1.61

[28] Martin Luther, *Luther's Works. Vol. 1. Lectures on Genesis*, ed. Janoslaw Pelikan (St. Louis, MO; Concordia Pub. House, 1958), pp. 30, 42-43.

solid firmament and the waters above the heavens! Ironically, the same folks who say that "any attack on the literal meaning of Genesis is an attack on the Gospel of Jesus Christ" are also guilty of dismissing key elements of the Hebrew cosmos as non-literal. Don't they understand the clear meaning of the text?

I'm not trying to pick on anybody, but all of this just goes to show you that no matter how hard Christians try to stay true to what we think the Bible says, it's practically impossible to keep modern scientific knowledge from influencing our understanding of Genesis. What was once so clearly *literal* to some is now dismissed as merely *figurative* by others—simply on the basis of scientific discovery. But if God's Word is unchanging and eternal, how does such an important element of the cosmos get demoted from *literal* to *figurative* in just a few hundred years? And what other significant events are subject to negotiation based on extra-biblical knowledge? Dr. Walton tells us how we can easily avoid these hermeneutical inconsistencies:

> The solution is to understand the worldview through which the text is communicating and focus on that which it seeks to communicate in isolation from our pro- or antiscientific agendas. We should not be asking (1) how the text validates my scientific understanding or (2) how the text describes the scientific system we know to the true; rather, we must ask (3) on what level the text is communicating its message. This does not require scientific apologetics and text manipulation; it requires comparative study.[29]

In other words, when we come across a biblical description of something like the ancient firmament, we shouldn't feel compelled to give it a modern scientific significance, nor should we make excuses for the text by putting words in the author's mouth; rather, we should just recognize it for what it is—a structural element of the ancient Near-Eastern cosmos that was commonly acknowledged by all primitive cultures.

This brings up an interesting dilemma for conservative Christians: how do we regulate the amount of extra-biblical knowledge that is allowed to inform our understanding of the Bible? We all do this to some extent, so how much is too much? Many well-meaning Christians claim that "only Scripture should be allowed to interpret Scripture" and nothing outside of the Bible should influence a straightforward reading of the text in question. This sounds all well and good, but do any of us still believe in the physical details of the

---

[29] John H. Walton, *The NIV Application Commentary: Genesis* (Grand Rapids, MI; Zondervan, 2001), pg. 94.

ancient Hebrew cosmos? Why not? Has the Bible changed since ancient times? Obviously not; but our scientific understanding of the cosmos certainly has changed—many times over. So if we are going to allow modern science to inform our interpretations of Scripture, then we had better stop pretending that we don't and instead start developing a consistent approach that doesn't sabotage the Christian faith by laying the entire Bible on the altars of naturalism.

Regardless of how we do this, one thing is clear: there is no going back. We simply know too much about the universe and how it works to return to the ancient cosmos. So where do we go from here? There is no question that the biblical firmament of Genesis was meant to be a literal solid structure supporting an ocean of water above the heavens, just as the "days" of creation were clearly meant to be 24-hour periods. Any interpretation of Scripture that tries to dismiss the solid firmament and the waters above is simply taking these verses out of context for the sole purpose of avoiding the "embarrassing" fact that neither of these things actually exists. But if we properly understand the actual point of Genesis, we shouldn't be embarrassed by the "clear meaning of the text." These passages can easily be explained in terms of the ancient Near-Eastern cosmogony that serves as the unmistakable framework of the creation narrative. By leaving these verses in their original context, we can avoid these potentially embarrassing situations that often force us to dismiss the text as merely figurative or symbolic.

Since Genesis appears to be a divinely inspired cosmogony contextualized for an ancient audience, modern science as we know it today is completely useless when trying to understand it—just like modern Hebrew and Greek probably wouldn't get us very far when translating the original biblical texts. For those of you who want to read Genesis as an accurate physical description of the modern cosmos, this has got to be disappointing news. Although the cosmos described in Genesis may have been true by the ancient standards of simple observation, it bears little resemblance to the universe as we know it today. The solid biblical firmament is just one example of this.

Why is Moses just regurgitating the conventional wisdom of Egypt and Mesopotamia? Was he simply falling back on his Egyptian education? Having a flashback from the "good old days" in Pharaoh's courts? Not so fast! To the Jewish people that had just left Egypt and were wandering in the desert on their way to the Promised Land, the Genesis account would have been incredibly significant. Consider their situation: after being oppressed and indoctrinated by the Egyptians for hundreds of years, there probably wasn't much reason to believe Yahweh was any match for their pagan gods. The Hebrews had just left the relative safety and security of the only life they had ever known, and were about to face the unknown hardships of life in the wilderness where they would constantly be exposed to the destructive forces of nature.

Having been liberated from the *order* of Egypt and thrust into the *chaos* of the desert (the original "out-of-the-frying-pan-and-into-the-fire" scenario), they longed for Egypt and her pagan gods—who were said to have created the heavens and the earth by organizing and restraining the very forces of chaos that threatened them in the wilderness. Even after God unleashed the plagues on Egypt—clearly demonstrating that Egypt's gods had no power over the forces of nature—the Israelites still complained to Moses, "Was it because there were no graves in Egypt that you brought us to the desert to die?...It would have been better for us to serve the Egyptians than to die in the desert!"[30]

What Moses brings down from Mt. Sinai elevates the God of Abraham, Isaac, and Jacob to unimaginable heights. Yahweh is in control. He alone restrains the forces of nature that threaten to destroy them. The universe is under God's command and He has established the boundaries of nature. By His voice, not the voice of Atum, God created the heavens and the earth. No longer were the pagan gods of their Egyptian and Mesopotamian oppressors given any legitimate status. The universe was borne of unity and harmony, not of violence and conflict. The biblical account shows how infinitely more powerful and purposeful the Hebrew God is than the self-serving quarreling deities of pagan mythology. In short, Genesis gives the ancient Hebrews a theology of creation that was much more profound than anything they would have been familiar with.

The Genesis account is exactly what the Hebrew people needed at that moment in history. To demand that Moses provide 21st century Christians with a scientific description of creation that meets our post-Enlightenment standards of material fact is the height of modern arrogance! Rather than give the Hebrews a new cosmology on their way out of Egypt, which would have been quite interesting to us but completely useless to them, God inspires Moses to reassure them—using language that they would have all been intimately familiar with—that Yahweh would watch over them en route to the Promised Land. Basically, the creation narrative was God's *theological* rebuttal to the Egyptian creation mythology, not a *scientific* rebuttal of ancient Near-Eastern cosmology.

I think we unintentionally marginalize the Scriptures when we dig through Genesis trying to piece together a literal description of the modern cosmos by finding proof-texts that can be supported by scientific evidence. If I had a nickel for every commentator that tried to explain the "waters above

---

[30] Exodus 14:11, 12
[31] Also repeated in Psalm 148:4 (NASB)

the heavens"[31] as water vapor in the form of clouds or ice crystals, I'd be rich. We can't read modern atmospheric science into the Bible and expect it to make sense! Even so, the clouds are not *above* the sun, moon, and stars so we just need to accept it and move on. Ironically, many of these same folks will sternly warn others about interpreting the six days of creation as anything other than literal days. Where is the consistency here?

This dogmatic approach to the natural sciences unfortunately plays right into the hands of the enemies of orthodox Christianity, who enthusiastically agree that the Scriptures are intended to teach creation science. But for atheists, deists, agnostics, and "liberal" Christians, the apparent scientific "errors" contained in the Scriptures only demonstrate their unreliability in matters of *both* science and doctrine. One of the many tragic consequences of the creation science movement is that it unintentionally undermines the essential Christian doctrines about God, man, sin, and the death and resurrection of Jesus Christ by casting doubt on the literal meaning of the entire Bible.

How many rational scientists have hardened their hearts towards spiritual things because they couldn't get past the apparent scientific errors of Genesis? I'd say there have probably been quite a few. Consider these words from a close friend and colleague of the late Carl Sagan (1934–1996):

> How was it, [Sagan] wondered, that the eternal and omniscient Creator described in the Bible could confidently assert so many fundamental misconceptions about Creation? Why would the God of the Scriptures be far less knowledgeable about nature than are we, newcomers, who have only just begun to study .he universe? He could not bring himself to overlook the Bible's formulation of a flat, six-thousand-year-old Earth...This newly acquired vision made the God who created *the* World seem hopelessly local and dated, bound to transparently human misperceptions and conceits of the past.[32]

This is very sad. How many others has Dr. Sagan led down this path of biblical rejection based on his own misunderstanding of the true intent of Scripture? But what I find even more troubling is that Dr. Sagan's premise, that God intended to provide mankind accurate scientific information in the pages of Scripture, is no different than the premise that underlies the creation science movement. In fact, Dr. Sagan probably took this disastrous cue from conservative Christians who thought they were just standing up for the Bible, when in fact they were unknowingly contributing to the hardening of hearts

---

[32] Carl Sagan, *The Varieties of Scientific Experience: A Personal View of the Search for God*, edited by Ann Druyan (New York, NY; Penguin, 2006), pg. x.

against it.

Christians need to understand the first chapter of Genesis for what it is: an "accurate" rendering of the physical universe *by ancient standards* that God used as the vehicle to deliver timeless theological truth to His people. We shouldn't try to make Genesis into something that it's not by dragging it through 3,500 years of scientific progress. When reading Genesis, Christians today need to transport themselves back to Mt. Sinai and leave their modern minds in the 21st century. If you only remember one thing from this chapter make it this: Genesis is not giving us creation science. It is giving us something much more profound and practical than that. Genesis is giving us a biblical theology of creation.

No Higher Criticism

Now at this point I have to clarify one thing. Ever since the ancient Near-Eastern creation mythologies were discovered by archeologists, the enemies of Christianity have tried to make the case that their obvious similarities to the Hebrew Scriptures prove that the Bible is not unique. Basically, these critics are accusing Moses of copyright infringement. The point of these attacks on the Bible is to try to make the case that Moses did not write under divine inspiration, but rather "off the top of his head" using unoriginal material.

Are these accusations fair? Why would we assume that Moses wrote Genesis in a cultural vacuum and didn't have access to the Hebrew language, the Hebrew figures of speech, the Hebrew customs, the Hebrew cosmogony, the Hebrew mythologies, and the Hebrew legends? These were simply the tools available to him at the time of his writing. That Moses seems to have used the popular ancient Near-Eastern creation mythology as the framework to develop a distinctly Hebrew theology of creation shouldn't weaken our faith in the least.

Trying to argue that the Bible is not inspired by God because it borrows from the popular mythology is like trying to argue that the Bible is not unique because is uses a pre-existing human language. Also consider this: before modern empirical science came onto the scene, mythology was the *medium* used by ancient cultures to express this kind of cosmic information. When God inspired Moses to replace the pagan polytheistic myths with monotheistic versions, why wouldn't he retain the basic non-religious elements of the popular mythology?

To even suggest something like this is offensive to most conservative Christians. Unfortunately, many of us have bought into the post-Enlightenment materialists' lie that all truth must be communicated in terms of newspaper-style journalism with identifiable sources, or in terms of a scientific dissertation with referenced footnotes. But what gives us the right to impose our 21st century Western standards of academic scholarship on an ancient Near-Eastern text? We have to be extremely careful here. Again I quote Dr. Walton:

> The very fact that the Bible's ability to use Israelite modes of think-ing poses such a problem for us demonstrates how significantly we have been influenced by certain aspects of our culture. We have been persuaded to believe that truth about origins can only be packaged in scientific terms; that the only cosmological reality is a scientifically informed reality; that if a cosmological text operates outside of the scientific realm, it ceases to be truth. We too easily accept the dictum that the only absolute is science. This presupposition causes us to think that the Bible's authority would be jeopardized if its revelation fails to address origins in terms that reflect our worldview.[33]

We should acknowledge that God, speaking through Moses, makes use of ancient mythology (the foolishness of the world) to write a distinctively He-brew mythology about Himself. And unlike the pagan myths that were writ-ten in ignorance of the one true God, the Hebrew myths accurately reveal the Living God to His chosen people using a medium that they would have read-ily identified, understood, and assimilated into their culture. What else would we expect God to write, an article for *Scientific American* complete with pie charts and color bar graphs? Or should He make a power-point presentation using fancy computer animation to illustrate every step of the creation process?

Now some Christians may have legitimate concerns over what this im-plies about the historicity of the Old Testament Scriptures, or at least the first eleven chapters of Genesis which share many elements of ancient Near-East-ern mythology. Once we start using words like "mythology" to characterize any portion of the Scriptures, it makes a lot of us understandably nervous. How do we know what stories are factual and what stories are fictional? Where do we draw the line?

These questions are understandable, but they are modern concerns, not ancient ones. If the purpose of ancient mythology as a medium of communi-cation was to show the character of the gods, to demonstrate how they con-tributed to the identity of a culture, and to explain how they gave order and purpose to the physical universe, then it seems entirely necessary that God

---

[33] John H. Walton, *The NIV Application Commentary: Genesis* (Grand Rapids, MI; Zondervan, 2001), pg. 89.

would begin the Bible with the only one true myth ever written; the *myth to end all myths* so to speak. It would be a shame to ignore the significance of the mythological context of creation just because some critics have tried to use it to cast doubt on the infallibility and inerrancy of the Bible.

So what then becomes of the fundamental doctrines of man, sin, covenant and redemption that all come from the Book of Beginnings? How can these things be valid unless every physical detail of Genesis is also assumed to be true? This is another understandable, but ultimately modern question. These precious doctrines are doctrines because God chose to make them doctrines by whatever means available, not because the stories that relate them to us as doctrines pass our modern standards of scientific and historical accuracy!

Nevertheless, the material details of the stories in question may be historically true, they may be loosely based on historical truth, or they may be more like the New Testament parables, but that should never be our primary concern. This is an interesting academic question perhaps, but to state that the truth of these doctrines completely hinges on the specific details of their historicity is to step out of the ancient mindset and into the mindset of a modern, western, post-Enlightenment materialist—an approach that would have been completely foreign to the generations that committed God's Word to writing.

Modern Flat-Earth Mythology

There can also be negative knee-jerk reactions by some Christians against any idea that people once believed the earth was flat based on the Bible. Actually, there is a very good reason for this response. During the 19th century, a concerted effort was made to characterize the first voyage of Christopher Columbus as a modern debate over the fundamental shape of the earth, which it clearly was not. The motivation behind this blatant historical revision was to try to paint a picture of medieval Christians as dogmatic idiots who would rather believe "falsehoods" told by the Bible than the "truth" as revealed by science. While the Scriptures do indeed make many passing references to the ancient flat-earth model of the cosmos, there is nothing within these texts that demands we adopt this model for our time. So any claim that belief in the inerrancy of Scripture also requires a return to the ancient Hebrew cosmos is simply ridiculous.

Nevertheless, part of me is sympathetic to this argument because the whole point of this book is to show how easy it is for Christians to misunderstand and misapply the Scriptures to the study of nature. In fact, medieval Christians had plenty of erroneous ideas about how the universe worked, some even based on the Bible, but a flat earth was not one of them. By that time, everybody knew the earth was round. Even the common people understood this. Yes, there were some cults who thought that the earth was still flat, but they were *pagan* cults. Nevertheless, the enemies of Christianity and the Bible have successfully transported the ancient flat-earth cosmogony thousands of years into the future and pinned it on the Bible-believing Christians of the late Middle Ages.[34]

Telling a lie just to make a point is not a very good way to win over your audience, even if your point is valid. As we'll see later in Chapter Five, by the time Columbus sailed across the Atlantic, many Christians were arguing from the Bible that the earth didn't move, and that the entire heavens revolved around it. The astronomers were offering scientific evidence to the contrary. If one wants to demonstrate the folly of taking certain passages from the Bible out of their ancient Near-Eastern cosmological context and using them to incorrectly establish scientific doctrines about the physical universe as we know it today, there are plenty of other interesting examples to work with. There is no reason to rewrite history just to try to prove a point. Let us not confuse the ancient Near-Eastern cosmogony that was universally believed by all Semitic peoples well before the time of Christ with the modern flat-earth myth invented by the enemies of Christianity.

---

[34] For a detailed explanation of how the modern flat-earth myth was fabricated by 19th century historical revisionists, see the book by Jeffrey Burton Russell, *Inventing the Flat Earth: Columbus and Modern Historians* (New York, NY; Praeger, 1991).

CHAPTER FOUR

MISSING THE POINT OF GENESIS

The next question to be asked is why would God intentionally recycle the erroneous cosmology of the ancient Near East? Actually, the real question is what would have been the point for God to overturn the established view of the physical universe? An incorrect view of the cosmos borne from ignorance is not a problem for God. I'm certain that even our best efforts today are lacking because of things that we just don't know. But an incorrect view of God borne from bad theology is a serious issue that must be addressed. That was obviously God's focus in Genesis. To put it simply, the intent of Genesis is theological, and any references of a "scientific" nature are culturally bound by the cosmology of the author and his audience.

The primary concern for the ancient Hebrews was *polytheism* versus *monotheism*, not the shape of the earth, the nature of the sky, the relative position of the stars and planets, or the biological origin of the species. Those are all issues that later generations of believers dealt with. When we force Genesis to address contemporary issues that would have been meaningless to the ancient Near-Eastern world, we actually undermine the authority of the Bible. How is that possible? Dr. John Walton gives us this warning:

> If we read levels of meaning into the text on our own that neither the audience nor the author would have understood or been concerned with, we set ourselves up as the channel for the text's authority. Since God's process of inspiration utilized the author, I believe the author's intention is our most important authority link. We must be willing to accept that it may not have been his concern to defend the ideas that our culture asks us to defend.[1]

God's obvious purpose in Genesis was to purge the universe of its many resident deities and establish Yahweh as the creator and sustainer of all things. The most direct way to accomplish this was to sanctify the existing cosmo-

---

[1] John H. Walton, *The NIV Application Commentary: Genesis* (Grand Rapids, MI; Zondervan, 2001), pg. 82.

logical framework and use it as the vehicle for theological reform. In His infinite wisdom, God knew that we would continue to refine our understanding of the physical cosmos through the process of discovery. In fact, by removing the gross polytheism of the pagan cosmogonies, the Genesis account gives mankind a rational basis for scientific naturalism and sets the stage for future scientific discovery. Not surprisingly, science has flourished in all monotheistic cultures throughout history. But as far as giving the Hebrews a new cosmos, Mt. Sinai was neither the time nor the place for that.

The question remains: why would God leave the ancient Near-Eastern cosmology intact? Any answer is only speculation, but sometimes it is easier to give new meaning to an old tradition than to convince people to forsake old traditions altogether. Take the Christmas and Easter holidays for example. Both of these traditions have their origins in the pagan feasts and festivals of ancient European peoples.[2] Each included the worship of false gods and goddesses. When the Christian Church started evangelizing the unreached regions of the European continent, it was very difficult to convince these people to forsake their pagan traditions. There was a real concern that bad theology would work its way into the church if people continued to observe these annual feasts. Theology was the primary concern, not ignorance. So in a brilliant tactical maneuver, the church hijacked both of these festivals and sanctified their meanings by relating them to significant Christian events: the incarnation and resurrection of Jesus Christ.

Rather than be forced to choose between celebrating these festivals and keeping their faith, the newly converted pagans could enjoy their customs and traditions and be reminded of essential spiritual truths at the same time. Not surprisingly, there is a small minority of Christians today that refuse to celebrate these holidays because neither is inherently Christian. That seems a little extreme to me, but certainly it is their right to do so. Ironically, these same people will probably fight you to the death over every last detail of the Genesis creation account, which appears to have its origins in pagan mythology as well. In light of the Christmas and Easter holidays, it seems very reasonable that God would sanctify the common creation tradition by firmly grounding it in theological truth.

But what would have happened if instead God gave the ancient Hebrew people a spherical earth and a solar system? What if they were told that the eight planets (sorry Pluto) of our solar system were tucked away in the suburbs of an ordinary galaxy, which was part of an ordinary galaxy cluster, and there were a billion others just like it? I can only speculate, but that probably wouldn't have been the best thing for them. After being oppressed for many

---

[2] Wade Cox, "The Origins of Christmas and Easter" http://www.logon.org/english/s/p235.html

generations, they were probably feeling a little insignificant in the eyes of God. Rather than boost their morale, a detailed description of the modern universe might have deflated them somewhat. By further removing God's chosen people from the physical center of His attention, it might have been more than they could have handled.

The Israelites were already a "stiff-necked" people. Their trust in God was about as fair-weather as the loyalty from the fans of a losing sports team. I think that a scientifically accurate picture of the cosmos would have been so alien to them that it could have ruined the already tenuous credibility that Moses had with the people. Or perhaps they would have stopped short of the Promised Land and started a university to teach the ignorant peoples of Mesopotamia and North Africa about the real universe? Nobody knows. Like I said, anything you come up with is just speculation. But fortunately it doesn't matter, because God didn't, and still doesn't work like that.

The Principle of Accommodation

God seems to be very patient when it comes to basic ignorance on the part of His children. Don't we do the same thing with our own children? If your toddler draws on the wall with a marker, you exercise restraint. You use the opportunity to instruct, to teach, and to instill principles and values that encourage respect for property, order, and cleanliness. After all, ignorance just means that he probably didn't know any better. Now if your teenager draws on the wall with a marker, that's just plain graffiti and is not tolerated. Why? Because he *should* know better. Ignorance is no excuse. Whether they were specifically forbidden to draw on the wall or not doesn't matter at that point. By the time you reach the teenage years you are expected to use your head.

At times, we might even contribute to our children's ignorance by contextualizing certain concepts that may be too difficult for them to handle at tender ages. We might refer to the passing of a loved one as "sleeping" or we may withhold the fact of adoption, or refer to sexual intercourse as some form of "advanced snuggling" for mommies and daddies. You probably have some creative interpretations of truth of your own. We don't always deal with our children strictly in accordance with the cold, hard facts. Instead we make allowances for their ignorance and immaturity because there are often greater issues that may be obscured by laying down the factual account. The bottom line is that we don't give them more than they can handle, even if it is the truth. And so our Heavenly Father often deals with us much in the same way.

We can see this clearly in the Christian doctrine of the incarnation. In the ultimate act of contextualization, God Almighty, maker of heaven and earth, eternal, omnipotent and omnipresent Lord, puts on human flesh, is born into a common family, fully embraces the human experience, and though He was without sin, He endured the ultimate suffering on our behalf. He did all of this for our sake, so that we might have a clear picture of the consequence of our sin and the infinite mercy of God. So if God can limit His very nature by entering time and space in the person of Jesus Christ, shedding His eternal and infinite attributes and voluntarily submitting Himself to His own creation, even to the point of death on a cross, certainly He has the artistic license to make use of the foolishness of popular mythology in order to contextualize the creation account so that the original non-scientific audience could receive it. This is known in theological circles as the *principle of accommodation.*

To put it another way, the drama of God's creation, man's fall, and Christ's redemption could have played out on any cosmological stage. But regardless of the actual cast, characters, props, and plot, the doctrines conveyed through that drama transcend each generation's concept of the physical cosmos. If a 21st century prophet were telling the creation story to a modern audience under divine inspiration, it would still start with the phrase, "In the beginning, God created the heavens and the earth." But after that, it would probably look different than the Genesis account. The theology wouldn't change, but the cosmological vehicle of its inspired delivery would be as familiar to us (even repeating the scientific "errors" of our time) as the ancient Near-Eastern cosmogony was to the Hebrews.

Think about this for a minute. If God were to directly reveal to us these timeless theological truths using any other cosmological model than our own, the central message would get lost in the technical details. The vehicle of delivery would completely *distract* us from the main point of the narrative rather than *enhance* it. Why do you think there is so much confusion over the meaning of Genesis today? Quite simply, the ancient *medium* of creation (which is mythological[3]) has distracted us from properly understanding the timeless *meaning* of creation.

This same principle of accommodation can be applied to any passage of Scripture that uses the ancient cosmos as a medium to communicate timeless theological truth. For example, Psalm 136:7 draws from the Genesis cosmogony when it talks about the sun and the moon as the "two great lights." Are we to infer from this that the moon is physically greater than the stars,

---

[3] Note: to say Genesis uses mythology as a medium of communication is not the same as saying that Genesis is a myth.

planets, and galaxies? Clearly not, but that is exactly how some Christians once understood this passage. While nothing in the text demands that we draw these astronomical conclusions, if we bring these kinds of nonsensical questions to the foothills of Mt. Sinai, we will get these kinds of nonsensical answers. In his commentary on Psalm 136:7, John Calvin (1509–1564) illustrates the principle of accommodation when he writes the following:

> The Holy Spirit had no intention to teach astronomy, and in proposing instruction meant to the common to the simplest and most uneducated persons, he made use by Moses and the other prophets of popular language, that none might shelter himself under the pretext of obscurity, as we will see men sometimes very readily pretended an incapacity to understand, when anything deep or recondite is submitted to their notice. Accordingly, as Saturn though bigger than the moon is not so to the eye owing to his greater distance, the Holy Spirit would rather speak childishly than unintelligibly to the humble and unlearned.[4]

This powerful principle of accommodation can have other applications as well. I recently read of another very practical example. The situation involved Western doctors trying to prevent the spread of infection by midwives in a primitive native culture.[5] Rather than try to teach them about bacteria and germs, concepts that had no familiar cultural context, the doctors decided to use the natives' own unscientific traditions to communicate the knowledge necessary for their "salvation." This instruction took the form of "ritual" washing so that "demons" from the hands of midwives will not be transferred to the baby or mother. The desired effect was achieved, even if by means of factually incomplete or incorrect knowledge.

Now ask yourself this: If these natives are ever to advance their knowledge to the point of understanding the actual material mechanisms by which infections are transmitted by unclean hands, will they curse these Western doctors for not giving them the factual truth? Or will they appreciate the wisdom of these doctors, accommodating their ignorance and meeting them in their time of need—so that despite their lack of knowledge, they might still be saved? What a wonderful picture of how God deals with us!

---

[4] John Calvin, *Commentary of the Book of Psalms*, trans. James Anderson (Grand Rapids, MI; Eerdmans, 1949), 5:184.

[5] Robin Collins, "Evolution and Original Sin," edited by Keith B. Miller, *Perspectives on an Evolving Creation* (Grand Rapids, MI; Eerdmans, 2003), pg. 478. The story was originally from Peter van Inwagen, *God Knowledge, and Mystery: Essays in Philosophical Theology* (Ithaca, NY; Cornell University Press, 1995), pp. 141-142.

The Hebrew Universe

It is clear from the rest of the Old Testament that the ancient Near-Eastern cosmos was fully adopted by the Jews. All of the biblical writers that followed Moses paid homage to it. Some Christians will initially take offense at this idea, but it is a perfectly reasonable assumption—considering that Israel had no way to know of an alternate cosmology unless God had directly revealed it to them. But there absolutely is no record of another ancient Near-Eastern cosmology. In fact, whenever the Bible makes a reference about the heavens and the earth, it fits pretty nicely with the Egyptian cosmos that Moses would have known so well growing up in the Pharaoh's courts.

You can see several references to the "four corners"[6] of the earth resting upon the "pillars of the earth"[7] which were laid upon the "foundations of the earth."[8] The sky itself is shown to be "stretched out"[9] like a tent and rests on the "pillars of heaven"[10] extending down to the "holy mountains."[11] Then of course the "firmament"[12] from Genesis is repeated and also the "waters above the heavens"[13] which are released when God opens and closes the "windows of heaven"[14] to bring rain upon the earth. There are even references to the "doors of heaven"[15] from which the sun enters and exits the sky each morning and evening. In fact, the Psalmist describes the sun as entering the firmament each day "like a bridegroom coming out of his chamber"[16] before making its "circuit to the other end"[17] of the firmament, presumably to return to its chamber.

This kind of language is repeated throughout the Old Testament, just as we would expect it to be. After all, this was the ancient cosmos, universally accepted by all people of that region during that time. It only makes sense that something written during that time would reflect the common understanding. The book of Enoch—while not officially part of the Bible we have today—provides a very detailed description of the Hebrew cosmos. Figure 2 illustrates the universe as the Hebrews would have known it.

---

[6] Isaiah 11:12; Proverbs 30:4 (NASB)
[7] Job 9:6; Psalm 75:3
[8] I Samuel 2:8; Psalm 104:5; Job 38:4-6; Jeremiah 31:37
[9] Jeremiah 10:12; Isaiah 40:22; Psalm 104:1-2
[10] Job 26:11
[11] Psalm 87:1
[12] Psalm 19:1; 150:1 (KJV); Amos 9:6 (NASB)
[13] Psalm 148:3-4 (KJV)
[14] Genesis 7:11; 8:2 (KJV)
[15] Psalm 78:23
[16] Psalm 19:5 (KJV)
[17] Psalm 19:6

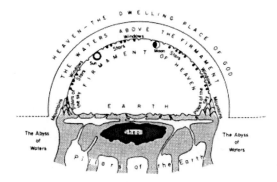

Figure 2: The ancient Hebrew cosmos as described by the Bible.[18]

Many Bible commentators dismiss these literal descriptions of the heavens and the earth as mere poetry, figurative language, or metaphor, but when taken literally they are actually pretty consistent with the ancient understanding of the physical universe. And why wouldn't they be? The biblical writers had no reason to break with the common understanding because no alternate cosmology had been revealed to them. Genesis 1 clearly reinforces the ancient view, giving it a sort of divine sanction. So in the context of the ancient Near-Eastern cosmos that was commonly understood during Old Testament times, these verses were actually very literal—all of them. They were literal for the age that committed them to writing. Does that mean they are wrong? Only if we ignore the principle of accommodation and incorrectly assume that they were written with intent to provide us accurate cosmological information in the first place.

By permitting the biblical authors to draw from their own experiences and knowledge, we don't have to make excuses for their lack of understanding. In his commentary on Genesis 1:6, John Calvin wrote, "nothing here is treated but of the visible form of the world..." If we take this approach, then the whole debate between a literal versus figurative interpretation of Genesis becomes somewhat a moot point. Given the non-scientific context of Genesis, it is perfectly natural for a modern person to hold to a literal *original* interpretation of the Old Testament without having to apply the physical details of the ancient cosmos to today's universe. It really just boils down to this: we don't need to force the Scriptures to fit the universe as we know it today because today's physical universe is of no concern to the Scriptures.

[18] James L. Christian, *Philosophy: An Introduction to the Art of Wondering, 6th ed.* (Harcourt, 1994), pg. 512.

I realize that not all readers will immediately embrace this approach, but what are the alternatives? If we boldly state that Genesis is some kind of coded language referring to a future cosmos that was unknown to both the author and the audience, then we are forced to reinterpret these Scriptures to fit the latest scientific discoveries. We will always be playing catch-up with scientific discovery. Moreover, once we start down that slippery and gradually sloping road, everything we hold sacred becomes subject to reinterpretation. We basically offer up the whole counsel of God on the altar of scientific naturalism. The tragic outcome to this approach is that the infallibility and inerrancy of the Bible is eventually called into question. There is simply no way the Scriptures can hold up under that kind of scientific scrutiny—not because the Bible is flawed, but because it was simply never intended to answer those kinds of questions. The biblical authors used only those things available to them at the time of their writing. These modern questions are entirely inappropriate.

The Six-Day Framework

The next logical question is what to do with the six-day framework in which the creation account was given? Were those literally six 24-hour days or were they symbolic of something else? There is no consensus among Christians. But of all the recent creationist claims, the idea of a young universe based on the six days of creation probably draws more secular fire than any other idea, so obviously we have to deal with it. But is it wise to use the biblical creation account to establish the age of the universe? Was that the original purpose of the narrative? Was the age of the earth an issue that Moses would have even addressed?

Certainly if we search the Bible for how long it took God to create the universe we will find an answer (more of this in Chapter Six), but is the question even valid? Do the material patterns of God's providence (the uniformity of nature) provide us with other ways by which we can piece together the natural history of the cosmos? What happens when the answer provided by Scripture is different than the answer provided by a scientific approach? I think we already know the answer to that question.

Given all that we've seen thus far, should the six-day framework even be a contentious issue? Why are we even stressing out over this? The Hebrew story of creation was obviously mapped onto the six-day Hebrew work week. The six days of creation are clearly six 24-hour periods of time. No matter how hard you try, you can't get around it. But neither can these six "days" be separated from the world that is created within them. And since that world was demonstrated to be physically inaccurate by subsequent discovery, the six-day framework really has no direct scientific application to a spherical earth traveling around a distant sun with seven other planets and a bunch of comets and asteroids. Nor should it have anything to do with the billions of stars in our own galaxy and the other billions of galaxies out there.

Here is another way to put it: if you take Genesis 1 and 2 as a literal play-by-play journalistic account of the construction of the heavens and the earth and add to it all of the Old Testament references to the ancient cosmos, the universe that you end up with is basically the same one that is shown in Figure 2. There is no way around this either; at least not without completely ignoring the plain meaning of the text. Furthermore, you can't just accept the six-day framework as scientific truth without also accepting the very same universe that it describes. And since the Hebrew universe is not an accurate scientific description of the *space* we now find ourselves in, then neither can the six-day framework be an accurate scientific description of the *time* we now find ourselves in. It simply was never intended to be. That wasn't the point of it. To even bring these kinds of questions to the Bible is nonsensical.

While many Christians automatically assume that the purpose of the creation week is to tell us how long it took God to make the cosmos, there is nothing in the biblical text itself that requires us to accept the literal creation week as scientific or historical truth. There certainly is an answer waiting for us there if we ask the question. But once again, is the question valid? Do we also assume that Psalm 135:7[19] is teaching meteorology, or that Psalm 103:3[20] is teaching medicine, or that Psalm 139:13[21] is teaching embryology? If so, then just about every field of modern science is in big trouble! Likewise, I don't see how we can assume that Genesis 1 was ever intended to communicate the scientific details of *how* God created the universe, unless we also assume that God knows less about the universe than we do. In each of these cases, the texts clearly do not claim to teach physical science, so any questions of a scientific nature are simply inappropriate.

Before moving on, let's look at an example from history specifically related to the creation week that shows what can happen when we ignore the ancient mythological context of Genesis in favor of a modern scientific interpretation. According to ancient cosmology, daylight was not necessarily a result of the sun. You might think this sounds ridiculous, but it wasn't that far fetched considering simple observation. Because the twilight appeared an hour or so before the sun entered the firmament through the "doors of heaven," the connection

---

[19] "He makes clouds rise from the ends of the earth; he sends lightning with the rain and brings out the wind from his storehouses."

[20] "...who forgives all your sins and heals all your diseases..."

[21] "For you created my inmost being; you knit me together in my mother's womb."

between the two was not firmly established. In other words, the sun was still in its "chambers" when the morning sky began to glow, so the twilight must have been caused by something else. This may seem like nonsense given what we now know about the solar system, but it was a perfectly logical argument when operating in the context of the ancient cosmos.

According to the six-day framework, the sun and moon were not created until the fourth day.[22] We often struggle with this fact because day and night were established by God on day one.[23] This passage sometimes gets us in trouble when others point out that for three days there were mornings without sunrises and evenings without sunsets. The reason it bothers us today is because we actually know how the solar system is supposed to work. We know how the sun's rays light up the atmosphere over the eastern sky before the sun itself can be seen over the horizon, so any suggestion that dawn and dusk are independent of the sun would be ridiculous to us. We also know that the force of gravity between the earth and sun is what holds the earth in its orbit. We can't conceive of the earth hurling through empty space apart from the sun, so the "clear meaning" of Genesis becomes somewhat of an embarrassing problem.

Unfortunately, any attempt to perform damage control by employing novel interpretations of the Scriptures only comes across as a desperate move to save face. Those who insist on a strict chronological application of the days of creation conclude that while the sun must have been created on the first day with the earth (for obvious scientific reasons), it was initially *hidden* from view and didn't become *visible* until the fourth day—probably because of cloud cover or a watery canopy. With enough crazy assumptions you can make just about interpretation fit the data (perhaps our levitating snow machines from Chapter One were obscuring the sun?), but the six-day chronology would not have been a problem for those holding to the ancient cosmology. Recall St. Ambrose, one of the early church fathers; he used the following argument against the Greek philosophers:

> We must remember that the light of day is one thing and the light of the sun, moon, and stars another—the sun by his rays appearing to add luster to the daylight. For before sunrise the day dawns, but is not in full refulgence, for the sun adds still further to its splendour.[24]

Can you imagine trying to use that line of reasoning today? You wouldn't even be taken seriously! The idea that day and night are independent of the sun is

---

[22] Genesis 1:14-19
[23] Genesis 1:3-5
[24] Hexameron, Lib. 4, Cap. III

absurd given the current model of the solar system, but these verses make perfect sense when left in their ancient Near-Eastern cosmological context. The six-day framework and the flat-earth cosmogony were simply never intended to give us that kind of scientific information about the universe as we know it today.

The End of the Road

We talked earlier about science having limits beyond which it can no longer provide reliable information, the "end of the road" so to speak. Well, here we see the limits of divine revelation beyond which it can no longer give us reliable information. The Bible simply can't provide us a scientific description of the modern physical universe. I know this is not going to sit well with many evangelicals, but you may as well ask a scientist to provide you with a chemical analysis of good and evil if you're going to search the Scriptures for the exact date of creation, or the distance to the stars, or the depth of the oceans, etc...

Once the Greek astronomers gave us a spherical earth, these "scientific" Bible verses became vestiges of the age that committed them to writing and we should always try to keep them in that context. There may be some passages here and there that appear to be consistent with the universe as we know it today. But for every one of those, there are a hundred others that will make no sense in a modern context.

It is very common for Christians to assume that, "The earth on which man lives today is the same earth whose six-day creation Genesis 1 relates..."[25] This might seem reasonable on the surface, but clearly it can not be the case. Where is the watery abyss? Where are the four corners of the earth? Where are the pillars of heaven? Where is the solid firmament? Where are the waters above the sun, moon, and stars? Where are the windows of heaven that regulate the rain? Where are the sun's chambers that it rests in each night? If we are going to backtrack by saying that these references are symbolic and figurative, then what about the six-day framework? Why are we to understand that literally, but not the solid firmament and the waters above it? And if we are free to assume that the firmament was figurative or symbolic of something else, then what about the plagues of Egypt, the Virgin Birth or the resurrection of Christ? What about the rest of the Bible?

Once we force Genesis to answer questions about the physical universe as we know it today, we unintentionally invite the entire scientific community to pick apart the biblical creation account. And they won't just stop at Gene-

---

[25] P. Andrew Sandlin, *Creation According to the Scriptures* (Valliceto; Chalcedon, 1991), pg. 3.

sis 1. The momentum gained from easily conquering the ancient Near-Eastern cosmogony will carry them through the entire Bible, tearing down one spiritual truth after another. By keeping Genesis in its ancient Near-Eastern context, we also keep it out of reach from science. The principle of accommodation allows us to unapologetically assert the *original* "clear meaning of the text" without putting the Scriptures at odds with the modern scientific rendering of the cosmos.

This doesn't mean that the Bible contains errors per se. I feel like I have to keep clarifying this point because many Christians will immediately jump to this conclusion and respond with a vigorous desire to "let God be true and every man a liar."[26] Perhaps you are feeling this way now. That's understandable, but rest assured that God's Word is always true and all men can and will be liars. However, we Christians are just as *human* as anybody else. We can also be "liars" by virtue of misreading and misrepresenting God's Word. Even worse, we can make God out to be a liar when our misguided claims are dismissed by simple observation.

If you have any doubt about this, just check out any material put out by the *Association for Biblical Astronomy*.[27] Your first thought should be, "Why do we need *biblical* astronomy—what is wrong with just plain astronomy?" But believe it or not, there are evangelical Christians today with PhDs in astronomy that still believe the earth is at the center of the universe and the entire heavens revolve around us. How can they believe this? Because the Bible clearly says that "the world is firmly established and cannot be moved."[28] And every biblical reference to a heavenly body refers to its "motion" across the sky and the earth is always said to rest firmly on its foundations. If we only use special revelation to solve this problem, then we are doomed because there is no biblical case for a solar system—and there are 67 verses that support the geocentric model. Moreover, the modern solar system was discovered through extra-biblical investigation, which is always held suspect.

Are questions about celestial mechanics even something we should ask the biblical authors? Or do we have other ways of finding out the answers to these types of questions? Were it not for the modern science of astronomy, people today would probably still believe that the earth is fixed and immovable. This doesn't mean we're all stupid, but how else would we know about the rotation and orbit of the earth when we can't even see it or feel it moving? A straightforward reading of the Bible only confirms what our senses already tell us—that the earth is at rest! This clearly demonstrates our need to under-

---

[26] Romans 3:4
[27] "The Association for Biblical Astronomy" http://www.geocentricity.com/
[28] Psalm 93:1

stand Scripture in light of both natural revelation and comparative literature. If we only used Scripture to interpret Scripture, we would all still be uncompromising geocentrists!

How Should We Then Read?

When we come across Bible verses that appear to give us scientific information, such as those telling us that the entire heavens revolve around an immovable earth, there are basically three options available to us: we can (1) take these verses as literal scientific truth and vigorously defend this model of the universe against all rival theories based on extra-biblical knowledge; or (2) take these verses as non-literal and reinterpret them in conformity with modern astronomy; or (3) understand these verses as giving us a literal, but non-scientific, view of the universe based on the popular cosmology of the age that committed them to writing.

The *Association for Biblical Astronomy* has basically chosen option (1), to only interpret Scripture with Scripture, and to place the testimony of the Bible in opposition to Copernicus, Galileo, and just about every other astronomer since the Renaissance. Is this a choice that we all must make: modern astronomy versus the Holy Spirit? As stated in their *geocentric primer*, "the issue is no less important than that of the authority of the Holy Bible."[29] Does this imply that the rest of us are just astronomical sellouts who are simply suppressing the truth in unrighteousness?[30] Should we also repent our sins, turn from our unbelieving ways, and join those who argue from the Scriptures that the sun, stars, and planets all revolve around the earth? Of course, most creation scientists do not believe this, but make no mistake; the *Association for Biblical Astronomy* has taken the "creation science" approach to understanding and applying Scripture to the study of nature.

Option (2) compromises the integrity of the entire Bible by allowing its readers to rationalize away any verse in which the literal meaning is contrary to modern understanding. This is a slippery slope that can lead to all kinds of hermeneutical abuses. Many times, the theological meaning of the passage in questions gets thrown out with the physical descriptions and the Bible becomes irrelevant to modern culture.

Again, I vote for option (3); to use the principle of accommodation. It preserves the integrity of the Scriptures without forcing unnatural applications of the ancient cosmogony to the universe as we know it today. This ap-

---

[29] Gerardus D. Buow, "A GEOCENTRICITY PRIMER: Introduction to Biblical Cosmology"; 2004. http://www.geocentricity.com/geocentricity/primer.pdf
[30] Romans 1:18

proach is quite liberating, giving Christians a respectable exit strategy from a pointless battle that we can't win.

How could we have missed this all of these years? Actually, this approach has not been entirely overlooked. There are plenty of examples in history of this very thing. Going back again to the earliest scientific controversies of the church, St. Augustine, in his commentary of Genesis, warned his fellow Christians against engaging in pointless scientific debates with educated persons using the Scriptures. He wrote the following:

> Usually, even a non-Christian knows something about the Earth, the heavens, and the other elements of this world, about the motion and orbit of the stars and even their size and relative positions, about the predictable eclipses of the sun and moon, the cycles of the years and seasons, about the kinds of animals, shrubs, stones, and so forth, and this knowledge he holds to as being certain from reason and experience.
>
> Now, it is a disgraceful and dangerous thing for an infidel to hear a Christian, presumably giving the meaning of Holy Scripture, talking nonsense on these topics; and we should take all means to prevent such an embarrassing situation, in which people show up vast ignorance in a Christian and laugh it to scorn. The shame is not so much that an ignorant individual is derided, but that people outside the household of the faith think our sacred writers held such opinions, and, to the great loss of those for whose salvation we toil, the writers of our Scripture are criticized and rejected as unlearned men....
>
> Reckless and incompetent expounders of Holy Scripture bring untold trouble and sorrow on their wiser brethren when they are caught in one of their mischievous false opinions and are taken to task by these who are not bound by the authority of our sacred books. For then, to defend their utterly foolish and obviously untrue statements, they will try to call upon Holy Scripture for proof and even recite from memory many passages which they think support their position, although they understand neither what they say nor the things about which they make assertion.[31]

Has anything really changed since this was written?[32]

---

[31] St. Augustine, *The Literal Meaning of Genesis*, translated and annotated by John Hammond Taylor, S.J., Vol 1. (New York, NY; Newman Press, 1982), pp. 42-43.

[32] At least one thing is now different: we have replaced the geocentric model with the solar system. Note how St. Augustine's reference to the "orbit of the stars" is consistent with biblical geocentricism, which was also the contemporary scientific view of his day.

Now fast-forward a thousand years. During the time of the Reformation, contemporary astronomy was shedding new light on medieval cosmological beliefs. When it was discovered that Saturn was a much greater object than the moon, but only appeared tiny because of its great distance, many Christians rejected this observation because the Bible clearly states that there are only "two great lights"—the sun and the moon. In fact, the stars were given a much "lower" status in Genesis when Moses writes, "…and He made the stars also," almost as if they were of inconsequential magnitude in relation to the sun and the moon. Clearly, a plain and straightforward reading of the Scriptures suggests that the moon is physically greater than the stars, entirely consistent with the scientific details of the ancient cosmos.

Once again, the Bible and common sense find themselves in complete agreement. Christians today don't have to try to explain this by putting words into Moses' mouth. All we have to do is recognize that the modern tools of astronomy, which often reveal truths that are contrary to common sense and the Bible, were not available to ancient cultures. I submit that were it not for modern science, we would all still believe today that the moon was a greater light than Saturn. How else would we know any better?

But not all Christians rejected this "new" astronomical discovery. John Calvin once again demonstrates the principle of accommodation by warning his fellow Christians against endorsing an astronomical position that was contrary to science just because it was described differently in the Scriptures. In fact, Calvin argues that we should allow for Moses to be "wrong" on astronomy since he did not have the benefit of contemporary observations to inform his scientific understanding. In his commentary on Genesis 1:16, Calvin writes the following:

> Moses makes two great luminaries; but astronomers prove, by conclusive reasons that the star of Saturn, which on account of its great distance, appears the least of all, is greater than the moon. Here lies the difference; Moses wrote in a popular style things which without instruction, all ordinary persons, endued with common sense, are able to understand; but astronomers investigate with great labor whatever the sagacity of the human mind can comprehend. Nevertheless, this study is not to be reprobated, nor this science to be condemned, because some frantic persons are wont boldly to reject whatever is unknown to them.[33]

33 John Calvin, *Commentaries on the First Book of Moses Called Genesis*, trans. John King (Grand Rapids, MI; Eerdmans, 1948), 1:85.

I find this statement very liberating. Calvin basically frees Christians from having to use these verses to establish scientific doctrine by allowing Moses to speak in simple non-scientific terms that ordinary simple-minded persons, "endued with common sense" can easily understand. He then endorses the idea that science is able to investigate nature on an entirely different level, and should not be condemned by "frantic persons" who reject things that they simply don't understand.

It was also believed, based on the fact that Moses calls the moon a "great light" that it possessed its own luminosity. But the first telescopes revealed that the moon had craters and mountains, clearly demonstrating that it was only reflecting the light of the sun. Despite this evidence, there were some Christians who rejected any idea that moon was not actually a "light" based on the words of Moses. Calvin concludes his commentary on Genesis 1:16 with the following statement:

> There is therefore no reason why janglers should deride the un-skill-fulness of Moses in making the moon the second luminary; for he does not call us up into heaven, he only proposes things which lie open before our eyes. Let the astronomers possess their more exalted knowledge; but, in the meantime, they who perceive by the moon the splendor of night, are convicted by its use of perverse ingratitude un-less they acknowledge the beneficence of God.[34]

Here we see Calvin basically instructing us to stop expecting the biblical authors to answer questions about modern astronomy. Elsewhere Calvin reinforces this idea when he says of Genesis that "nothing is here treated of but the visible form of the world. He who would learn astronomy, and other recondite arts, let him go elsewhere."[35] Now think about that statement for a minute. Given the careful attention that Calvin paid to contemporary questions of astronomy, it is obvious that the hot-button issues of his day were astronomical. But can we also take this same approach to the hot-button issues of our day, like biology or geology? Can we not also say of Genesis, "He who would learn *geology* or *biology*, let him go elsewhere"? Is that what Calvin had in mind by the phrase, "other recondite arts"?

In the mid-1800s, there was another creation science movement attempting to reconcile contemporary geology with the biblical details of creation and the Noahic flood. In response to the "Scriptural Geologists," Harvard botanist and Congregationalist Asa Gray (1810–1888) said the following:

---

[34] Ibid, 1:85-86.
[35] Ibid, 1:79.

We may take it to be the accepted idea that the Mosaic books were not handed down to us for our instruction in scientific knowledge, and that it is our duty to ground our scientific beliefs upon observation and inference, unmixed with considerations of a different order...I trust that the veneration rightly due to the Old Testament is not impaired by the ascertaining that the Mosaic is not an original but a compiled cosmogony. Its glory is that while its materials were the earlier property of the race, they were in this record purged of polytheism and nature-worship, and impregnated with ideas which we suppose the world will never outgrow. For its fundamental note is, the declaration of one god, maker of heaven and earth, and of all things, visible and invisible, - a declaration which, if physical science is unable to establish, it is equally unable to overthrow.[36]

In addition to recognizing the ancient Near-Eastern context of the Genesis creation week, Gray also realized that if we keep the Bible from making scientific claims that can be overturned by observation, then we also must give up the search for empirical proofs that God is the maker of heaven and earth, and of all things, visible and invisible.

A Greater Witness

Some Christians may not think this approach is any big deal, as they have always believed by faith and they don't look to science to confirm their beliefs. But unfortunately, so much of the Creation Science movement is laser-focused on apologetics, as if the only thing the unbelieving world needs before they can put their trust in Christ is scientific evidence for God and creation. These tactics clearly do not consider the fallen human condition. Unbelievers do not reject God for lack of evidence, but for lack of faith.

Paul tells us that "For since the creation of the world God's invisible qualities—his eternal power and divine nature—have been clearly seen, being understood from what has been made, so that men are without excuse."[37] For someone who has hardened their heart towards spiritual things, there is no evidence, scientific or otherwise, sufficient to soften it. Even if they were to directly witness a bona fide miracle, it would only further harden their heart and prompt the immediate search for a rational explanation. At that point, piling

---

[36] Asa Gray, *Natural Science and Religion: Two Lectures Delivered to the Theological School of Yale College* (New York, NY; Scribner's, 1880), pp. 6-9.

[37] Romans 1:20

on what we think are clever scientific proofs for the claims of Christianity isn't going to get us anywhere. Creation Science is not able to break down these walls of unbelief and soften the hearts of the unsaved.

When Christians give up on Creation Science, scientific "proofs" for the existence of God and the authority of Scripture become unnecessary. I know that this doesn't sit well with our modern empirical sensibilities, but is this really so bad? Should scientific creationism even be a part of our Christian witness in the first place? Is this the kind of thing that missionaries study before traveling the globe to reach distant people for the Kingdom of God? Were the great periods of revival throughout church history the result of clever scientific arguments for God's existence?

CHAPTER FIVE

LEARNING FROM OUR MISTAKES

C hurch history didn't wait too long before providing us with many colorful examples of what can happen when Christians take the creation science approach to biblical interpretation.[1] The science of astronomy was advancing rapidly in Greek culture. New discoveries were challenging the church's acceptance of the ancient cosmogony. In every case, some in the church answered these challenges by combing through the Scriptures to find support for a scientific view that would eventually crumble under the weight of opposing physical evidence. Every time this happened, Christians were forced to retreat, their influence was weakened, and their authority was diminished. These early controversies between Christianity and science clearly demonstrate the futility of trying to piece together a scientific description of the physical universe using the Scriptures.

The Sins of Our Fathers

In the 6th century B.C. the Greek mathematician Pythagoras (c.580–c.500 B.C.) reasoned from mathematical principles that the earth must have a spherical shape. He had little physical evidence for this, but in mathematics the sphere represented the perfect form and efficiency of a three-dimensional object, so he concluded that the earth and all other heavenly bodies must take that form. Plato (c.428–c.348 B.C.) also adopted this view after studying Pythagorean mathematics. His favorite student, Aristotle (c.384–c.322 B.C.), used limited physical observations to support the emerging theory of a spherical earth. Finally in 240 B.C. the issue was settled scientifically when Eratosthenes (c.276–c.194 B.C.) measured the circumference of the earth within 2% of its actual value by comparing the an-

---

[1] Many of the references from this chapter are based on the research of Andrew D. White (1832-1918) who studied the historic conflicts between science and religion. White was known to exaggerate the conflict by overstating the case. Most of the examples he cited are not disputed, but many historians agree that they did not always represent the majority view within the church. The examples given in this chapter are simply to illustrate how the Scriptures can be wrongly interpreted, regardless of whether they were majority or minority opinions within the church at the time.

gles of shadows at different locations on the earth. And so proof of a spherical earth was firmly established roughly 270 years before the Christian Church had its first communion service.

The majority of Christendom accepted this discovery without reservation, but the comments of some early church fathers on this matter should encourage Christians today to exercise a little more humility toward scientific discovery. Rather than consider the evidence that the ancient cosmogony may be incorrect, some insisted that the biblical references to the old system were absolute, and could not be challenged. To them, the Bible was more than sufficient to provide answers to the details of creation and the evidence showing the spherical earth was not taken seriously. Contrast this attitude with the Greek gentiles who had no theological reason to hold on to the ancient theory and readily accepted the spherical earth based on the evidence. This should caution us today about being so quick to commit our interpretations of the Scriptures to scientific ideas that can eventually be disproved by natural revelation.

By the 2nd century, some church fathers such as Theophilus of Antioch (d. A.D. 182) were responding to this "new" theory by renewing their allegiance to the ancient cosmogony.[2] They searched through the Scriptures for any texts they could find making reference to the "foundations of the earth" and the "corners of the earth" and the "pillars of heaven" and the "windows of heaven" and the "waters above the firmament" and constructed an elaborate defense clearly showing the biblical support for the ancient theory. For some Christians, the idea of a spherical earth, popular with many "infidels" outside the church, could not be separated from the anti-Christian philosophies of its supporters. Rather than examine the physical evidence for the spherical earth as a separate issue, it was summarily opposed by some as anti-Christian heresy.

In the 3rd century, Clement of Alexandria (d. A.D. 216) further elaborated on the ancient theory and by the 6th century, the Egyptian monk Cosmas Indicopleustes developed an incredibly detailed description of the ancient cosmogony based on the construction specifications of the Jewish tabernacle. From the dimensions of the table to the number of loaves of bread and lamp stands in the tabernacle, Cosmas was able to "determine" the exact measurements for many of the universe's features. He had angels pushing and pulling the heavenly bodies across the firmament and even included a detailed doctrine for the regulation of the rain that also required angels to open and close the "windows of heaven" on God's command. Being Egyptian, it's no surprise that his universe resembled a vault with a domed canopy rather than a plane-

---

[2] It was actually over 400 years old by this time.

tarium. He concluded this detailed description by stating,

> We say therefore with Isaiah that the heavens embracing the universe
> is a vault, with Job that it is joined to the earth, and with Moses that
> the length of the earth is greater than its breadth.[3]

While the shape of the earth was still being debated, another controversy emerged. If the earth was really spherical, as some said it was, could the underside be inhabited by people? Some reasoned that the lands on the other side of the earth, which were called antipodes, could in fact be inhabited by people. Many of the church fathers strongly rejected this assertion as contrary to Scripture. In the 4th century, Lactantius (A.D. 245–325), who still did not accept that the earth was a sphere, formally opposed the habitability of the antipodes with the following statement:

> Is there any one so senseless as to believe that there are men whose
> footsteps are higher than their heads?...that the crops and trees grow
> downward?...that the rains and snow and hail fall upward toward the
> earth?...I am at a loss what to say of those who, when they have once
> erred, steadily persevere in their folly and defend one vain thing by
> another.[4]

This "common sense" logic was supposed to show that the idea of a spherical earth was ridiculous because the idea of habitable antipodes was ridiculous, hence the "one vain thing by another" comment. Arguing against the existence of antipodes on philosophical grounds is excusable, but the issue should have never been elevated to a theological debate involving Scripture.

As is usually the case, once Christians bring the full force of the Scriptures to bear on an argument, no evidence or argument to the contrary is even considered. I've been guilty of this exact same thing and many readers probably have too. Any scientific idea that contradicts the literal meaning of the Bible must obviously be a result of the natural man suppressing the truth in unrighteousness.[5]

Our hero St. Augustine, the great Bishop of Hippo, being one of the more reasonable leaders of the early church, was open to the idea of a spherical

---

[3] Andrew D. White, *A History of the Warfare of Science and Theology in Christendom* (New York; Appleton, 1898), Chapter I. Even by the standards of his time, Cosmas was considered a "kook" by most of his Christian contemporaries.

[4] Ibid, Chapter II. For other than cosmological reasons, Lactantius was considered a heretic by many in the church.

[5] Romans 1:18-20

earth. However, his reason for rejecting the idea of habitable antipodes was that "Scripture speaks of no such descendants of Adam."[6] He also argued that since Paul declared in Romans "their voice has gone out into all the earth, and their words to the ends of the world,"[7] and no missionaries had yet preached the Gospel at the antipodes, then the antipodes couldn't exist. Case closed.

Procopius of Gaza (A.D. 465–528) declared in the 6th century that:

> If there be men on the other side of the earth, Christ must have gone there and suffered a second time to save them; and, therefore, that there must have been there, as necessary preliminaries to his coming, a duplicate Eden, Adam, serpent, and deluge.[8]

So basically, there can't be people on the other side of the earth because the Bible never mentions such people being descended from Adam or being redeemed by the preaching of the Gospel. We are then left with a difficult choice: either there is another Adam and another Christ for the inhabitants of the antipodes, or there are no inhabitants of the antipodes, or perhaps there are no antipodes because the earth is actually flat. And that leads to my next question: would this second Christ of the antipodes be the antichrist? It's in the Bible, isn't it? And does the Jesus "down under" have an Australian accent? But seriously, before you just dismiss these arguments as ridiculous relics of the past, I submit that some of the logic we use today against modern science is just as bad when you break it down.

This same line of reasoning continued well into the age of Columbus (1451–1506), even though by that time the sphericity of the earth was almost universally accepted within the church. Even after Magellan (1480–1521) discovered the inhabitants of the antipodes during his voyage in 1519 to South America, the idea was still opposed by a small minority of Christians for two more centuries.

Extra-Terrestrial Evangelism

Is there anybody today who would argue from the Scriptures that the Aussies, the South Africans, the South Pacific Islanders or the South Americans can't be saved? Does anybody still think that there was another Adam, another Eve, another serpent, another flood, another crucifixion, and another resurrection for those people down under? Of course not! So what do you

---

[6] Ibid, Chapter II.
[7] Romans 10:18
[8] Ibid, Chapter II.

think happened between then and now? Has the Bible changed? Have the requirements for salvation changed? No! The world has become a smaller and more accessible place and our attitudes about the salvation of "distant" people have changed. Or have they really?

Now I just can't help but draw some parallels from the antipode controversy to our current day. I'm not seriously into UFOs nor do I have any hopes that we'll ever discover intelligent life outside of our solar system, but whenever I hear those in the church argue so dogmatically against even the possibility, I can't help but see the antipodes argument all over again. If there is intelligent life out there, it's not up to us to figure out how God might redeem those creatures, or if that is even necessary. And we have no business dismissing the possibility of finding them just because the Bible doesn't mention them.

The Bible doesn't mention extraterrestrials because there was no such thing as outer space in the ancient cosmology! It's the same reason it doesn't mention the Aboriginal peoples of Australia, because there was no "Down Under" in the ancient cosmology either, unless of course you were talking about the fires of hell or the fountains of the deep. I'm going to keep repeating myself on this point. The concept of the universe that is woven into the Scriptures does not exist today. Therefore, there is no scientific information in the Scriptures that we can fit into the modern universe.

Galileo and Copernicus

The eventual acceptance of a spherical earth by the church led to a new sacred theory of the universe. This was called the Ptolemaic system. Claudius Ptolemaeus (A.D. 83–161) was a Greek speaking Roman astronomer who lived at the turn of the 1st century. Even though he lived before the church formally accepted the spherical earth, the geocentric model of the universe that replaced the ancient cosmogony was named after him.

The Ptolemaic system placed the spherical earth firmly in the center of the universe, surrounded by ten concentric spheres, eight of which controlled the motions of the heavenly bodies. The closest sphere contained the moon (1), the next one Mercury (2), then Venus (3), the Sun (4), Mars (5), Jupiter (6), Saturn (7) and finally the stars (8). The ninth sphere was called the *Primum Mobile*, which imparts motion to the other spheres as the prime mover. The tenth sphere was called the *Empyrean* and was the absolute edge of the universe. Beyond that was the highest heaven, home to God and His angels. The Ptolemaic system (sometimes called the *geocentric model* because the earth, not the sun, was at the center) was widely accepted during the Middle Ages. Figure 3 illustrates the Ptolemaic system.

The geocentric model represented the best that science had to offer during the time that it was firmly held. It was entirely consistent with both naked-

Figure 3: The Ptolemaic System (public domain)

eye observation and philosophy. It was equally accepted and endorsed by both science and religion. The problem is that while scientific conclusions are always tentative, the Christian Church—just as some did with the ancient cosmogony—decided to build an elaborate theological and scriptural defense of the geocentric model. By failing to apply the lessons of the past, the church once again foolishly committed itself to a popular scientific theory supposedly based on the testimony of the Scriptures. The following statement is typical of the Christian position during that time.

> Just as man is made for the sake of God—that is, that he may serve Him,—so the universe is made for the sake of man—that is, that it may serve *him*; therefore is man placed at the middle point of the universe, that he may both serve and be served.[9]

What gives us the right to draw these kinds of *scientific* conclusions from a

---

[9] Ibid, Chapter III.

theological premise? To take a timeless spiritual truth about man's unique relationship to God and necessarily conclude that the earth must be fixed at the
center of the universe puts the entire Word of God in danger of being falsified
by simple observation!

As the Middle Ages were drawing to a close, a Polish astronomer named
Nicholas Copernicus (1473–1543) suggested in 1514 that perhaps the earth
and the other planets all revolved around the sun. At that time, the Catholic
Church was preoccupied with the Protestant Reformation and little attention
was initially given to this novel idea. Copernicus continued to refine his heliocentric theory until his death in 1573, publishing a book entitled, *On the
Revolutions of the Heavenly Spheres.*

The Italian astronomer Galileo Galilei (1564–1642) was heavily influenced by the claims of Copernicus. Early in his career, Galileo was challenged by opponents of the Copernican theory that "If your doctrines were
true, Venus would show phases like the moon."[10] Galileo agreed, and in 1611
his newly invented telescope observed just that very phenomenon. In his
mind, the Copernican theory was proven true.

Galileo traveled to Rome in 1612, just as opposition started to organize
against the ideas of Copernicus. He continued to gather evidence for this idea
and in 1632 he published a book called, *Dialogue Concerning the Two Chief
World Systems.* This book compared the evidence for the Ptolemaic and
Copernican theories and suggested that only the Copernican model could ad-
equately explain all of the observations. This publication basically laid waste
to the physical inaccuracies of the Ptolemaic system. But since the church had
elevated the geocentric model to the level of religious doctrine, Galileo represented a serious threat to the church's authority.

One of the claims of Galileo was that Jupiter had its own moons. He observed these heavenly bodies with his telescope as early as 1610, yet surprisingly this idea was opposed by some Christians. On what basis? By no less
than the testimony of Holy Scripture. You see, according to some, the seven
golden candlesticks of the Apocalypse, the seven candlesticks of the Jewish
tabernacle, and the original seven churches of Asia were proof that there
could only be seven heavenly bodies![11]

Another claim of Galileo was that the moon had mountains and valleys
casting shadows on its surface indicating that it was being illuminated by the
sun. Again, many Christians disputed his claim. On what basis? By the testimony of the Holy Scriptures of course. Genesis 1:16 describes the moon as a
"light" and therefore it clearly possesses its own luminosity. So any ideas

---

[10] Ibid, Chapter III.
[11] Uranus, Neptune, and Pluto had not yet been discovered.

about the moon reflecting the light of the sun are contrary to sound doctrine!

Pope Paul V, after examining the teachings of Galileo issued the following decree:

> The first proposition, that the sun is the centre and does not revolve about the earth, is foolish, absurd, false in theology, and heretical, because expressly contrary to Holy Scripture. The second proposition, that the earth is not the centre but revolves about the sun, is absurd, false in philosophy, and from a theological point of view at least, opposed to the true faith.[12]

The nature of the evidence that was given against Galileo was both scriptural and philosophical. The scriptural evidence shouldn't really bother us today if we remind ourselves that the Bible is based on the ancient cosmogony and always describes the earth as fixed and the sun and stars as moving across the firmament. It should come as no surprise that we can find plenty of verses that are contrary to the heliocentric theory. Once again, since hindsight is always 20/20, almost nobody still takes those verses as scientific fact so I'm not going spend any time explaining them. But the philosophical evidence given against Galileo is so comical—given the current model of the solar system—that it would be a shame not to look at it.

The Catholic Church actually declared that if the Copernican theory were true,

> The wind would constantly blow from the east.

> Buildings and the earth itself would fly off with such a rapid motion that men would have to be provided with claws like cats to enable them to hold fast to the earth's surface.

It was also asserted by others in the church that:

> Animals, which move, have limbs and muscles; the earth has no limbs or muscles, therefore it does not move. It is angels who make Saturn, Jupiter, the sun, etc., turn round. If the earth revolves, it must also have an angel in the centre to set it in motion; but only devils live there; it would therefore be a devil who would impart motion to the earth...

---

[12] Ibid, Chapter III.

The planets, the sun, the fixed stars, all belong to one species –
namely, that of stars. It seems, therefore, to be a grievous wrong to
place the earth, which is a sink of impurity, among these heavenly
bodies, which are pure and divine things.

If we concede the motion of the earth, why is it that an arrow shot
into the air falls back to the same spot, while the earth and all things
on it have in the meantime moved very rapidly toward the east?[13]

I'm not making this stuff up folks! These are real examples of what can happen when science is held hostage by our misinterpretations of the Scriptures!

Now just in case any of you Protestants think that this was just a problem
for the Catholic Church, consider the words of the great Reformers. We all
love Martin Luther (1483–1546), but he made the following unfortunate
statement about Copernicus:

People give ear to an upstart astrologer who strove to show that the
earth revolves, not the heavens or the firmament, the sun and the
moon. Whoever wishes to appear clever must devise some new system, which of all systems is of course the very best. This fool wishes
to reverse the entire science of astronomy; but sacred Scripture tells
us that Joshua commanded the sun to stand still, and not the earth.[14]

Philipp Melanchthon (1497–1560), another prominent Lutheran Reformer, in
his treatise on the *Elements of Physics* had this to say:

The eyes are witnesses that the heavens revolve in the space of
twenty-four hours. But certain men, either from the love of novelty,
or to make a display of ingenuity, have concluded that the earth
moves; and they maintain that neither the eighth sphere nor the sun
revolves…Now, it is a want of honesty and decency to assert such no-
tions publicly, and the example is pernicious. It is the part of a good
mind to accept the truth as revealed by God and to acquiesce in it.[15]

Even John Calvin, whom I've quoted as an example of how to properly approach Genesis, still believed in Geocentricism. In the introduction to his
Commentary on Genesis, he wrote the following:

---

[13] Ibid, Chapter III.

[14] Martin Luther, *Table Talk*.

[15] Andrew D. White, *A History of the Warfare of Science and Theology in Christendom* (New York;
Appleton, 1898), Chapter III.

We indeed are not ignorant, that the circuit of the heavens is finite, and that the earth, like a little globe, is placed in the center.[16]

So apparently we've found something that Catholic and Protestant traditions were united on: neither understood the proper relationship between science and the Bible! They both failed to see the seemingly cosmological verses of the Bible in the proper context and ignored the lessons of the previous flat-earth controversy.

It wasn't until the early 1700s when the scientific evidence of the "double motion of the earth" became so overwhelming that nobody in the church could still defend against it. Eventually, the heliocentric model was fully embraced by Christians. By the mid-1700s, the great Renaissance astronomers Copernicus and Galileo, who were condemned as heretics by the inquisition, were finally exonerated by the Catholic Church.[17]

The Cycle Repeats Itself

And here we conclude the second great battle between science and religion, and also the second embarrassing retreat for the Christian Church. In both of these hard-fought battles, the scientific theories survived no worse for the wear and the church was eventually overwhelmed by the facts of the case and had to recant in light of the scientific claims. In both cases, the principle of accommodation would never have put the Bible at odds with modern astronomy, and these conflicts may have been entirely avoided.

Both of these historic showdowns between science and religion equally demonstrate the folly of the creation science approach to biblical interpretation, but the rhetoric over the battle with Copernicus and Galileo was particularly shrill. The stakes seemed higher the second time around. Why was that? What was so offensive to Christians about the heliocentric theory? What's the big deal if the earth is just a planet revolving around the sun?

I think I have an idea. It's just speculation of course, but I think it has to do with our perceived place in the universe. Think about it. For thousands of years the entire universe fit into a sphere with roughly the same diameter as the earth we know today. Humanity, the crowning achievement of God's creation, was at the center of this tiny universe. The Ptolemaic system gave us a

---

[16] John Calvin, *Commentary On Genesis*, Edited and Translated by John King, M.D., (Carlisle, PA; Banner of Truth and Trust, 1975), Volume I, Part I.

[17] It was not until 1992 that Galileo was officially rehabilitated by Pope John Paul II.

spherical earth, but it was still firmly fixed at the center of the universe, immovable and absolute. The ten spheres of heaven encircled the earth. The earth was the stage, in the heavens were the spotlights and humanity was the main attraction. We were like rock stars—the headlining band in the concert of creation! And like the literal rock stars of our day, the entire universe revolved around us!

I'm sure there was a certain sense of security in this. It seemed reasonable to believe that if man was indeed the crowning achievement of God's creation, he should occupy the center of the known universe. So in light of this, I can see why the idea of earth as merely a planet revolving around a distant sun would have been so offensive. It cuts into our deeply held desire to be important. It's the same reason why we get so angry when someone cuts us off in traffic. Any challenge to a person's concept of self-worth and self-importance can definitely lead to irrational behavior—like road rage.

Combine the earth's motion around the sun with the fact that the other planets had a similar motion and it would seem as though man's special place in the universe was ruined forever. The idea that mankind is so spatially insignificant was contrary to everything that Christians believed about themselves. So the heliocentric theory touched a nerve, and defensive behavior naturally followed.

Now fast-forward to the present. What is it that really offends us today? What are the scientific discoveries threatening our sense of self-worth, our sense of self-importance? We seem to have come to terms with our spatial insignificance. Astronomers figure that the universe extends about 12 billion light years in every direction. It probably goes farther than that but the light from those regions hasn't even had time to reach us yet so we can't see them.

Most people seem to be OK with this now. Ironically, the whole idea now gives us a sense of wonder and amazement at the universe that God created. We took something that was a serious threat to our collective ego, turned the other cheek, and used it to our spiritual advantage. Creation books everywhere now proclaim the vastness of space as a testament to the wonders of the universe and the power of God to organize it all! We can now rejoice with David by saying, "What is man, that thou art mindful of him...?"[18]

So then what is it today that insults us the way the heliocentric universe insulted the Christians of the past? I think that what bothers us is the idea of *temporal insignificance*. What do I mean by that? One thing that is so offensive to Christians about scientific theories like evolution or the Big Bang is that these ideas place us so far toward the end of the timeline, that humanity just seems like an afterthought in the mind of God.

---

[18] Psalm 8:4 (KJV)

Take the latest cosmological predictions of a universe that is just under 14 billion years old and an earth that is about 4.5 billion years old. That means the universe existed for over 9 billion years before a cloud of dust and gas, probably the remnants of dead exploding stars, coalesced into the solar system we have today. That puts our planet in the last third of the universe. If all of history were a 60-minute game of hockey, our planet doesn't even get to play until the 3rd period!

Combine that with the latest theory of biological evolution that places the beginning of life on earth about 3.8 billion years ago with the first appearance of our species in the fossil record about 200 thousand years ago. If all of that is true, then I'd say humanity is about as insignificant in time as we are in space. Going back to our hockey game, how would you feel if the coach didn't even put you in the game until the last 49 milliseconds! Are you offended by this? Or will you also turn the other cheek? Perhaps someday creation books everywhere will marvel at the vastness of time and use it as a testimony to the wonders of creation and the power of God.

The whole point of this little detour through church history has been to show how using the Bible to create detailed theories about the physical universe hasn't worked out too well for us. In fact, it's been pretty embarrassing, to the point where our faith isn't taken seriously by many scientists anymore. Somehow along the way we've managed to isolate some of the most brilliant minds of our time from the Christian faith. And what else have we to show for this approach to science? What has it accomplished?

Has creation science cured any diseases? Extended the longevity and quality of human life? Reduced the infant mortality rate? Brought affordable housing to third world countries? Has creation science discovered anything significant that can be used in a science textbook to teach anything other than creation science? Can somebody please tell me one positive thing that creation science has done for us?

On the other hand, regular old science has accomplished plenty of amazing things. Why do some Christians feel that they need their own version of science? Do we feel the same way about other scientific disciplines? Take meteorology for example:

> To affirm that God is the Lord of clouds, lightning, rain, and wind (see Ps. 135:6-7) does not demand, or even suggest to me, that I should deny the possibility that the patterned behavior of matter serves as the proximate cause for all of these meteorological phenomena. To affirm the reality of God's governance of atmospheric phenomena does not make the science of meteorology unnecessary or un-Christian. The Christian meteorologist has just as much need

of the atmospheric sciences as does the naturalistic meteorologist. We have no need for specifically "Christian meteorology," using biblical barometers and theistic thermometers and angelic anemometers and regenerate radar instruments; we simply need good, honest, religiously neutral meteorology.[19]

And the same can be said for astronomy, cosmology, biology, geology, paleontology, and zoology.

The universe we find ourselves in today gives us a very different creation story than the biblical narrative does, one that has left many clues for us to uncover—if only we have "eyes to see" and "ears to hear." The people of God have come a long way since Mt. Sinai. As 21st century Christians, we have many more tools at our disposal, giving us the ability to develop a scientific rendering of the ancient creation story, keeping the theological truths intact while updating what was the ancient method of their divinely inspired delivery. In fact, the Creator has allowed us to uncover many of its mysteries, little by little. Look how far we've already come since the "underwater planetarium" of ancient mythology. The people of God, guided by a biblical theology of creation that unifies the natural sciences under the providence of God, should be leading this effort rather than using the Scriptures to oppose it!

Remember that God doesn't change, His nature and character do not change, His Word doesn't change, but our understanding of the universe has, and it may do so again. So the infallibility and inerrancy of the Holy Scriptures *demand* that we understand it in its original context. If we fail to do this, we unintentionally make God out to be a liar. It's that simple. As much as I hate to summarize myself, most of what I've said so far can be captured in the following bullets:

- Creation Science *removes* the Bible <u>out from under the protection</u> of the ancient Near-Eastern worldview—where truth did not always have to be expressed in scientific terms—and *subjects* it to the rigors of the modern materialists' worldview, which requires truth to always be precisely communicated in scientific terms.

- Creation Science marginalizes the *timeless* theological realities of the creation story by exposing the *timely* physical details to scientific ridicule.

- Creation Science hands the secular world a biblical straw man that can be easily torn down, along with the entire Christian faith.

---

[19] Howard J. Van Till, *The Fourth Day: What the Bible and the Heavens are Telling Us About the Creation* (Grand Rapids, MI; Eerdmans, 1986), pg. 235.

- Since all scientific propositions are subject to falsification, Creation Science puts the Lord our God to the test!

A Theology of Creation

At this point in our journey together, it is quite possible that you have a lot of conflicting thoughts racing through your head. You might even feel a little depression mixed with a tinge of betrayal, like finding out that the grandfather who you always admired as an example of integrity was actually a habitual womanizer. Clearly, the church's reaction to the claims of Copernicus and Galileo was just as embarrassing as the religious Crusades, American institutional slavery, and the Salem witch trials. To think that we could possibly be heading down this same path today is probably an uncomfortable feeling.

As a fellow traveler on this confusing journey, I know exactly how you are feeling. On the one hand, you want to stay true to the Scriptures. You want to take seriously what the Bible has to say about creation and you want to have a biblical approach to the natural sciences that honors the Creator. I actually couldn't agree more. On the other hand, you might be thinking that we probably shouldn't keep making scientific claims based solely on the Bible if it was never intended to provide modern audiences with detailed scientific information in the first place. But if that is true, are we to then conclude that science is an inherently secular discipline that has nothing to do with God and the Bible? What about *Sola Scriptura*? What about a comprehensive biblical worldview? What is a distinctively Christian approach to science supposed to look like if we can't even use the Bible as a scientific text?

These are all legitimate concerns, but what if a biblical approach to science didn't simply reduce us to finding physical evidence consistent with certain passages of Scripture just to demonstrate a tenuous connection between science and religion that constantly needs to be updated? This approach, characterized by the modern creation science movement, is nothing more than a dangerous deal with the devil—granting science exclusive rights to define reality on its own terms as long as its conclusions agree with what *we think* the Bible is telling us. The scientific method is then held hostage by predetermined non-scientific conclusions, undermining the entire enterprise and unnecessarily drawing Christians into a never-ending struggle with the scientific community over how to interpret the data.

Most Christians don't even realize that the relationship between science and the Bible is infinitely more profound than that! A truly biblical approach to science means that our scientific presuppositions must be firmly rooted in the Scriptures. In other words, what Christians need is a consistent *theology of creation*, not a distinctively *Christian version of science*. Hopefully by now you are willing to consider that constructing detailed theories about the cosmos by "proof-texting" biblical passages completely misses the theological point of the Genesis creation account.

Part of the problem is that modern Christians tend to have a compartmentalized view of life. We've adopted a dualistic view of the universe that divides everything between the secular and the sacred. We don't see how meteorology, medicine, or mathematics have anything to do with God so we give those subjects over to "the world" and declare a holy monopoly on things like politics, ethics and creation. We are perfectly content to treat most of the natural sciences as secular endeavors until, of course, somebody starts asking questions about our natural origins. Then, all of the sudden, we feel like the discussion has crossed over into "sacred territory" and we dig out our Bibles from under the stack of mail and bills and search for life's ultimate answers. Does this seem like a consistent approach? If we are truly seeking a comprehensive biblical worldview, it should probably look different than this.

Don't misunderstand my point here. The problem is not that we pull our Bibles out "on cue" as soon as certain key questions are asked; the problem is that we put them away in the first place. If we truly had a consistent biblical view of everything from astrophysics to zoology, we wouldn't feel compelled to overcompensate for our general abandonment of the sciences by recklessly applying certain Bible verses to the scientific investigation of natural history without first understanding what nature itself is telling us. Like those in St. Augustine's day, we often find ourselves on the wrong side of the obvious, giving the unbelieving world one more reason not to take the Bible seriously.

As Christians, we should always start with a proper view of nature. A biblical theology of creation teaches us that God created space, time, and matter from nothing. Consequently, nothing exists apart from God and He upholds and sustains all things. Nature does not act independently. It does not have a will of its own and does not govern itself. The uniformity and regularity that we observe in nature is a direct result of divine governance, or providence. These material "patterns of providence" manifest themselves as the laws of nature. Because these laws are *discernable* and *universal*, they provide a rational basis for mankind to study the world around him. Because nature is not itself God, but depends on God for its very existence, we can investigate *all* natural phenomena without fear of trespassing on Holy Ground.

There is, however, an important catch. We need to understand that a biblical theology of creation *is not* just another falsifiable claim that can be scientifically proved or disproved—it is not merely "creation science" or "scientific creationism." We don't need to waste time and energy trying to support it with scientific evidence because the theology of creation *is* the very foundation of scientific naturalism. That also makes it the overarching presupposition that unifies the natural sciences.

Does that qualify as a *Christian* approach to science? I think it does. So how about we all just acknowledge this fact and get on with the business of honest science without fear of "whoring after other gods" in the process? In my opinion, these distinctions between "secular science" and "Christian science" are entirely artificial. In fact, you wouldn't be wrong to say that all of science is inherently *theistic*, since it presupposes the created order based on a biblical theology of creation. Now I'm not saying that we should make much adieu about this and strut around the laboratory like we own the place, but certainly that is the proper perspective.[20]

A Skeptical Orthodoxy?

Many readers might still be uncomfortable with this, but what if Christian scientists, like other scientists, were free to follow the evidence wherever it leads them rather than force the data to conform to a predetermined outcome based on a misunderstanding of Scripture? And if an objective scientific investigation into the natural history of the cosmos just so happens to demonstrate beyond a reasonable doubt that the universe was, in fact, created over a period of six literal days some 6,000 years ago in a fully developed state—then shame on us for ever doubting the divinely inspired details of creation in the first place! In this case, it might be safe to conclude that perhaps Moses really was giving us accurate scientific information about the formational history of the physical universe. But really, is a little skepticism all that bad?

Upon hearing the news of the bodily resurrection of Christ, the Apostle Thomas said, "Unless I see the nail marks in his hands and put my finger where the nails were, and put my hand into his side, I will not believe it."[21] And rightly so; for the Apostle Paul tells us that if Christ were not raised, then our faith is futile![22] I don't know about you, but I don't have time for a "futile" faith. So indeed, if Christ did not rise from the dead, we are nothing but members of an irrational cult; no different than the fraternity of idiots who claim to have been fondled by space aliens aboard flying saucers.

Thomas may have been the only one to actually verbalize it, but I'm sure the others were having similar doubts. After all, it takes a lot of faith to believe that someone you just saw hanging dead on a cross can be made alive again. Everybody has to figure out what their own personal "threshold of faith" is going to be. Like the skeptical Apostle, I say that unless I see the ev-

---

[20] From this point on I'll drop the "secular" qualifier and just use the term "science" to refer to mainstream science. If I'm referring specifically to the scientific work of creationists, I'll still qualify it by using the term "creation science."

[21] John 20:25

[22] 1 Corinthians 15:17

idence for a flat and immovable earth, a solid firmament holding back an ocean above the heavens, and a young cosmos with only a few thousand years of natural history under its belt, I will continue to assume that the Bible was never intended to provide scientific answers to those types of questions. INERRANCY AND INFALLIBILITY DEMAND NO LESS OF US!

From God's Word to God's Works

The next part of the book takes an honest look at the creation story as told by the creation itself. If we can truly view the natural sciences as gifts of God based on the material coherence of the created order, then we should be free to draw objective scientific conclusions about natural history without fear of losing our faith. For Christians who wisely reject the presuppositions of creation science, having a "biblical" approach to creation simply means that we approach the study of nature with a different set of presuppositions, like those listed here.

Don't just quickly read through them, but actually consider each one carefully. These assumptions are much more practical than the assumptions of creation science because they don't tie the Bible down to any particular conclusions that force strange interpretations of Scripture or require questionable scientific theories.

- Nature is not something to be worshiped and it does not have a will or act on its own volition.

- The creation, being called into existence from nothing by the Word of the Lord, is entirely dependent on God to continuously uphold and sustain it.

- The laws of nature arise from the discernable patterns of God's divine providence, but are not absolute.

- These perfect "patterns of providence" were wisely ordained by God as the proximate cause of all ordinary natural phenomena.

- The continuous operation of the laws of nature is entirely sufficient to achieve God's creative purposes in the fullness of time.

- The omnipotence and omniscience of the Creator do not require (or preclude) that the material coherence of nature be miraculously interrupted.

- The laws of nature, being both discernable and continuous, are able to reveal the material mechanisms by which God's creative purposes unfold over time.

How would these assumptions change our approach to the study of nature? Well stay tuned, because that is exactly what the rest of the book about.

PART III:

WHAT CAN NATURE TELL US ABOUT ITSELF?

CHAPTER SIX

THE APPARENT AGE OF THE UNIVERSE

C reationists and evolutionists do have one thing in common: they both agree that the universe had a beginning. Beyond that, the two sides don't seem to agree on much. Despite their many differences, the significance of this little piece of common ground shouldn't be overlooked. Because the first book of the Bible opens with the phrase, "In the beginning..." many faith traditions believe that *time* had a beginning, and it was at this *instant* when all things came into existence. But in every generation from Aristotle to Einstein, science had assumed that the universe was *static* and *eternal*—just as it had always appeared to be.

The assumption of a "steady state" universe was also convenient for those seeking to avoid any discussion about the beginning of time. After all, if time has a beginning, then "who" or "what" *began* it? And why is there something instead of nothing? Questions like these tend to make scientists uncomfortable because (1) modern man insists that all truth be described in scientific terms, and (2) science has no way to answer these kinds of questions. Moreover, without any physical evidence to support a creation event, science had no reason to believe otherwise.

All of that started to change in the 1920s. The revolution of modern physics that began in the 1900s was spilling over into the science of astronomy giving stargazers the ability to ponder the heavens on a whole new level. Advances in telescope technology enabled astronomers to collect an increasing amount of data from the heavens. These data were kicked around for almost forty years until a "new" idea about the origins of the cosmos emerged. By the end of the 1960s, most scientists were finally ready to agree with Moses that the universe really does have a beginning! One astrophysicist sums up the situation like this:

> For the scientist who has lived by his faith in the power of reason, the story ends like a bad dream. He has scaled the mountains of ignorance; he is about to conquer the highest peak; as he pulls himself over the final rock, he is greeted by a band of theologians who have been sitting there for centuries.[1]

---

[1] Robert Jastrow, *God and the Astronomers* (New York, NY; W.W. Norton, 1992), pg. 210.

Now before we all start holding hands, singing Kumbayah and building extra pews for all of the scientists who will be lining up outside our churches, there are some things you should probably know about the scientific version of the creation story. For instance, the universe is not understood to be merely a few thousand years old as many in the church have suggested, but most scientists estimate that the age of the universe surpasses several billions of years.

If you're like me, you've probably been conditioned to react negatively to any claim that the universe is billions and billions of years old. The whole idea of that much time passing before Adam and Eve walked in the Garden kind of gives us the creeps. I'll be the first to admit that the first time I looked at just exactly what it is that compels so many people to believe in evolution, I felt like I was betraying my faith. Like the feeling that you get after you've just sat through a movie that you know you shouldn't have seen (patiently waiting through two hours of foul language for that "redemptive" sub-plot that must have ended up on the editing room floor). But if you've been track-ing with me through the book so far, and you're willing to adjust your expec-tations of both special and natural revelation, you may also be ready to at least find out what all of the fuss is about when it comes to these scientific creation theories.

Science as a Mission Field

To agree or disagree with the mainstream scientific creation theories as being ultimately *right* or *wrong* is really not the point here. Not to say that sci-entific claims are unimportant—because they are—but if science can't make absolute claims about reality, how can they ultimately be *right* or *wrong*? Now this is just my opinion so don't take it to the bank, but terms like *right* and *wrong* probably shouldn't even be used, in an ultimate sense, to evaluate sci-entific claims—simply because they cause us to overlook the tentative nature of scientific knowledge. During the Middle Ages, one could have said that the geocentric model was "right"—for all practical purposes. Since people back then were not concerned with launching satellites into orbit, the geocentric model was quite useful. But over time, there was an increasing amount of data that couldn't be explained by the geocentric model. As a result, the heliocen-tric model became "right" and the geocentric model became "wrong". But once it was discovered that our solar system did not occupy the center of the Milky Way galaxy, the heliocentric model joined the "wrong" club—even though it is still used in limited applications. Our current understanding is that the universe has no actual center. I guess that makes the *acentric* model "right" and all previous versions "wrong."

The important point here is that when discussing a particular scientific model, what really matters is its *usefulness*—how well it accounts for the known data or how useful it is for making reliable predictions. In fact, scientific models that are technically "wrong" can still be "right" for specific applications. For example, we still use the geocentric model when describing the "motion" of the stars and planets from the reference frame of earth. We all still say "sunrise" and "sunset" without fear of sounding ignorant. When others look at a planet through my telescope, I tell them to look quickly before it "moves" out of view. If I told them to "look quickly before the earth moves" it just wouldn't sound right. The heliocentric model is not very *useful* for communicating that kind of information.

So the focus of this discussion is simply to understand what the secular creation theories are all about, to address the philosophical motivations behind them, to appreciate why so many intelligent people believe them, and to not fear them or feel threatened by them. If we are really serious about having constructive dialogue with evolutionists, then we should first put ourselves in their shoes and understand why they hold the positions that they do. This might also help us to understand why secular scientists react so strongly against most creationist claims and thus avoid unnecessarily isolating this group from the Gospel message.

Think of it like this: say you want to be a missionary to another culture. This culture has different beliefs, different myths, and legends about who they are and where they came from, different traditions and different customs. If you show up with nothing but your Western cultural baggage and expect to win friends and influence people, you'll probably just end up in a pot of boiling water. If you want to have any chance with these people, you first need to learn their language, understand their beliefs, know their myths and legends, respect their traditions and honor their customs. You need to be willing to contextualize the Gospel message so that it can be received and assimilated into their culture. You need to be able to employ the principle of accommodation. Your goal as an ambassador of Jesus Christ is not to replace other cultural forms with Western Christian cultural forms, but to redeem all cultures by the power of the Gospel. After accepting Christ, each culture will take many of their traditions and customs, the things that are not inherently contrary to Scripture, and give them new theological meaning consistent with biblical Christianity. The final result will be a unique expression of Christianity that looks different *in form* than the one we're familiar with, but has the same theological *substance* as any other orthodox community of faith.

Now consider the secular scientific community as another culture to be redeemed by the Gospel. That's not too far-fetched. So how well have Christians done in this area? How successful has the dialogue with the scientific community been? I'd say not very. A survey conducted in 1998 of members of the United States National Academy of Sciences showed that less than 7% of its members believe in a personal God.[2] I'm sure it hasn't gotten any better since then and probably much less than 7% even belong to a local church. If this isn't a cultural mission field right here in our own backyard then I don't know what is. What language do scientists speak? What do they mean when they talk about the *Big Bang* or *cosmic evolution* or *evolutionary biology*? What are the motivations and philosophies that shape this culture? How can we present the Gospel message to them so that the only offense is the Cross itself? Do we really need to raise the bar of saving faith to the point where an honest scientist has to recant his or her entire life's work in order to accept the Bible as God's Word?[3] Which of their beliefs, traditions, and cultural forms will take on new theological meaning after being redeemed by Christ?

If we are serious about taking a *missional* approach to scientific culture, then these are the questions we must ask. I know it can be difficult because I've been there myself. Culturally, many of us are very uncomfortable when discussing naturalistic theories about the origin of the universe. We have a tendency to reject scientific descriptions when they fail to relate the operation of the physical cosmos directly to God's transcendent purposes. When natural phenomena are described as "random" or "undirected" (which could have a *theological* or a *scientific* meaning) we are prone to take theological offense—completely forgetting about the inability of science to draw transcendent conclusions about the nature of reality. Rather than recognize that some physical systems are so complex that they defy our ability make accurate predictions (like the weather), we read theological significance into words like "chaos" and dismiss the idea on spiritual grounds—knowing that not even a sparrow falls to the ground apart from the providence of God.[4]

The writing style of some scientists can also create confusion. Some writers, in a noble effort to make scientific literature more inspiring than the user's manual for your computer's operating system, tend to anthropomorphize the cosmos. A statement like, "organic molecules spontaneously organize themselves into large spherical structures" might imply that some intelligence apart from God resides deep within the properties of matter itself. Now this could be the result of a pantheistic philosophy (not very common in the scientific community), or it could just be a botched attempt at making the cosmos "come alive" by dressing up science with elements of mythology. These misunderstandings are unfortunate, but if we first acknowledge that science

---

[2] Edward J. Larson and Larry Whitham, "Scientists and Religion in America" *Scientific American*, September, 1999.

[3] Galileo had to recant everything before the Inquisition or face severe punishment.

[4] Matthew 10:29

is incapable of addressing transcendent concerns, then we shouldn't take offense when science intentionally avoids or unintentionally mishandles them.

We can diffuse these explosive situations by first making the appropriate adjustments in our expectations of a text according to the medium being used to communicate the intended information. If we fail to do this, we can either miss the point of the text completely, or we can read things into the text that were never intended by the author. For instance, when I read the story of Paul Bunyan to my children, they do not come away thinking that the Grand Canyon was literally created by a giant lumberjack who dragged his axe across the desert. Nor do they dismiss the entire story as foolishness because the material details are historically and scientifically inaccurate. Rather, they immediately recognize the mythological medium being used and adjust their expectations accordingly. This enables them to look beyond the incredible material details and allows them to focus on the larger themes that the text intends to communicate—like the strength and vitality of the American frontiersmen who "conquered" the Wild West by bringing order out of chaos (where have we heard that theme before?). Along these same lines, when people of faith examine secular scientific accounts of creation or natural history, they must adjust their expectations accordingly and not be disappointed when these texts ignore or bungle the religious implications. In fact, we should expect it!

The Age Debate

Let's get right down to it. The best estimates of modern science put the universe at about 13.7 billion years old. That's a lot of years. Written out, it looks like 13,700,000,000. When you compare that to the traditional biblical estimates, there appears to be a significant disagreement. The biblical data, like any other data, have some uncertainties that leave room for interpretation, but most leaders in the Creation Science movement don't think that the biblical data, if taken literally, will support a date of creation before about 10,000 B.C.[5] The estimate that I see most often puts the date of creation at 4004 B.C.[6] So as I write this now, the biblical age of the universe is anywhere from 6,000 to 12,000 years.

You don't have to be a math major to see that these two estimates are not even close to one another. How can two groups of people look at the same data and come to two entirely different conclusions? The answer is they can't. These are not just two different interpretations of the same data, but these are two different interpretations of two entirely different sets of data: there is the

---

[5] Henry Morris, *The Genesis Record* (Grand Rapids, MI; Baker, 1976), pg. 45.
[6] As determined by the Archbishop James Ussher in the year 1640.

biblical data and there is the astrophysical data. Which one are Christians supposed to believe?

The biblical data that gives us a 6,000 year-old universe is buried within the Old Testament genealogies. If you've ever tried to read through the Bible, but couldn't get past the parts that say "so and so begat so and so, who lived to the age of such and such, and begat so and so," then you probably ran across one of the genealogies. It may not have seemed that important to you at the time, and if you're like me you probably skipped over it, but in ancient cultures family connections were a very big deal. So it shouldn't surprise us that there are a lot of genealogies recorded in the Old Testament.

What makes these genealogies so important spiritually is that they show the bloodline of Christ from Adam, through Noah, Abraham, and David, in accordance with the Old Testament prophesies and covenant promises of God. The obvious focus of the biblical genealogies is prophetic and theological, but they are detailed enough in terms of chronology that when combined with a good historical knowledge of the ancient kingdoms, they can get you from Adam to Christ in about 4,004 years, give or take. And since the heavens and the earth were literally created in a span of six 24-hour days, the time between Adam and Jesus, plus the time since, pretty much gives you the absolute age of the physical universe.

So why is there such a huge disparity between the biblical data and the astrophysical data? Is it possible that we've missed something in our interpretation of Scripture? Can natural revelation be that far out of sync with special revelation? Or is this just another sad case of modern science "suppressing the truth in unrighteousness"? These are all great questions, but the real question is why would the universe *appear* to have a 13.7 billion-year natural history if the Bible tells us that it's actually very young? What's interesting about this question is that most Christian apologists will readily admit that the universe does in fact *appear* to have a very long natural history, much longer than a few thousand years. How do I know this? It's simple. Just look at the popular creation science literature, especially the Young-Earth oriented material. Nearly all of the creation science arguments are focused on explaining why the universe is not really as old as it looks. You can summarize these arguments with a statement like, "things aren't really as they seem and here is why..." So if things aren't really as they seem, then why do they *seem* like something else in the first place?

A Static or Expanding Universe?

Since the time of Newton and the dominance of classical physics, science has been uncomfortable with the longstanding assumption of an eternal universe. The reason is quite simple. Think of the entire universe as just two galaxies. According to the law of gravity, these two structures, no matter how large or small and no matter how far apart, will attract one another. No matter how insignificant this force is between them, these two structures will eventually collide if given enough time. So if the universe was eternal, then everything would have collided already and formed a supermassive black hole. The exact same thing would happen to a universe with three galaxies, four galaxies, five galaxies, or billions of galaxies like our own. Therefore, no matter how stable the universe looks from earth, it can't be static. And if the universe is not static, then it can't be eternal either. Without some other force that opposes gravity, everything would have collapsed under its own weight an eternity ago, leaving a massive black hole. Since that obviously hasn't happened, the age of the universe must be finite—hence there must have been a definite "beginning." If just thinking about these kinds of things gives you headache, you're not alone.

One way to make the universe eternal and still avoid this inevitable collision is to assume that it possesses some kind of unknown force that opposes the collective gravitational force. In 1917, Albert Einstein did just that when he developed some equations to describe the universe based on his general theory of relativity.[7] Much to his surprise, these equations showed that the universe could not just be static, but had to be either expanding or contracting. At the time, there was little observational evidence[8] for an expanding or contracting universe, so Einstein arbitrarily took one of the terms of his cosmological equations that had been previously been set to zero, a placeholder for an unknown repulsive force, and gave it a value that allowed the universe to remain static and eternal. Problem solved.[9]

This ad hoc term became known as the *cosmological constant* and Einstein would later refer to it as the "greatest blunder of my life."[10] I actually think he was being a little hard on himself. While his original cosmological equations did show an expanding universe, you can't really blame him for deciding to rework them to agree with the traditional view. Going out on a limb

---

[7] Einstein's general theory of relativity published in 1916 defines gravity in terms of geometry, space and time.

[8] I only say "little" because an astronomer named Vesto Slipher of the Lowell Observatory in Flagstaff, Arizona had previously observed some evidence for an expanding universe, but it usually takes a "preponderance of evidence" to overturn a longstanding popular scientific belief.

[9] In 1922 the Russian Physicist Alexander Friedmann demonstrated that Einstein's Cosmological Equations would lead to an unstable universe. Einstein eventually agreed with him and retracted his equations.

[10] Ironically after the discovery of *dark energy* in 1998, scientists found that the cosmological constant really is *nonzero*. This mysterious force is thought to be responsible for the increasing rate of acceleration of cosmological expansion.

with a bold claim having nothing more than some equations would be a gutsy move, even for a genius like Einstein.[11] One of the harsh realities of science is that it usually takes more than a just theoretical calculation to overturn a longstanding assumption. It also takes data. And the more revolutionary the claim, the more data you need to substantiate it.

During the 1920s, technology was enabling astronomers to look deeper and more carefully into space. They were not only making more detailed observations, but they were gathering more accurate data as well. One of the significant discoveries during this time was known as *galactic redshift*. That means that light waves coming from distant galaxies, when measured by sophisticated equipment, appeared to be stretched out toward the red end of the spectrum. There is only one thing that is known to cause this phenomenon and you probably already know all about it—especially if you're a fan of NASCAR races.

Something very similar to redshift happens to sound waves when a noisy object speeds by.[12] Since the speed of a sound wave is constant through air[13] and it can't be sped up or slowed down by moving the source, a rapidly approaching car squeezes its own sound waves together making a higher pitch, but a rapidly receding car stretches the sound waves out making a lower pitch. This squeezing and stretching produces the characteristic, eeeee-ooooo sound when your favorite driver speeds by.

Given everything we know about physics, stretched out light waves from distant objects can only mean one thing: these galaxies are speeding away from us![14] Nearby galaxies didn't seem to have this problem. Light waves from some of our nearest extragalactic neighbors, such as the Andromeda galaxy, appear to be squeezed together rather than stretched out. These light waves are shifted toward the violet end of the spectrum, indicating that these objects are slowly moving towards us because of local gravitational attraction. But once we get outside of our own local galaxy cluster, it becomes clear that other galaxy clusters seem to be moving away from our local cluster, and they appear to moving away from one another as well, indicating that the entire

---

[11] Of course, that is exactly what he did when he published the special and general theories of relativity in 1905 and 1916 respectively. So perhaps Einstein had some other philosophical attachment to a static universe that prevented him from initially opposing it.

[12] I use the word "similar" because the cosmological redshift is not entirely analogous to the Doppler Effect. The Doppler Effect assumes that your source is moving *through* space. Cosmological redshift occurs because space *itself* is expanding.

[13] It's actually dependant on the temperature, atmospheric pressure, humidity and density of the air, but unless any of these parameters are rapidly and drastically changing while a sound is traveling, we can pretty much treat it as a constant.

[14] Again, the "motion" is not absolute motion but a function of the expansion of space between us and them.

universe is being stretched out, or expanding. This idea might initially seem strange, but that's just what the heavens are telling us.

You may be surprised to know that this news was not immediately well received. Anytime a longstanding view of the universe is challenged, the defenders of the old system will hold out as long as they can, until the evidence becomes overwhelming. But just as the flat-earth model of the ancient world gave way to the geocentric model of the Middle Ages, which gave way to the heliocentric model of the Renaissance, the eternal static universe would eventually be replaced with an expanding universe that has a birthday.

By the end of the 1920s, the American astronomer Edwin Hubble had collected enough data to accurately measure the actual expansion rate of the universe. He did this by comparing the distance and speed for hundreds of galaxies and found there was a striking correlation.[15] The data seemed to fit the new model almost perfectly. With the rate of expansion determined, it seemed like all that was left to do was roll back the cosmic clock until everything squeezed back together and, poof! There you have the apparent age of the universe!

Of course, this is a lot easier said than done. Modern physics was still brand spanking new when Hubble first measured the apparent rate of cosmic expansion. His original estimation of a two billion year-old universe was based on an assumption that the rate of cosmic expansion was fairly constant. What makes the analysis so challenging is that the universe was much hotter when it was smaller. And when it was really small and really hot, things would have also been really weird—too weird for the regular laws of physics. The only way for physicists to really know how matter would have behaved in the primordial cosmos is to recreate similar conditions in a laboratory and observe what happens. The quest for this knowledge started a whole new science called *high-energy particle physics.*

The machines that do this kind of thing are called *particle accelerators,* and physicists kept building bigger and bigger ones. These things can take the same tiny particles that were believed to have made up the early universe and hurl them into one another at speeds almost as fast as light, generating tremendous amounts of heat and energy. The results of these experiments give us hard data on how ordinary matter behaves under the extreme conditions of the hot, small, primordial cosmos.

Talk to any high-energy particle physicist and they will probably tell you that bigger and more power particle accelerators are still needed, and perhaps they are. But I'd say that these guys have still been pretty successful at un-

---

[15] Astronomers now have data for over 20,000 galaxies to supplement Hubble's initial observations, further refining the Hubble constant.

covering the strange behavior of matter in the early universe. Just how successful exactly? Well, they claim to know enough about high-energy physics to recreate the conditions going back to the first 0.00000000001 seconds of time! Before that, I'm sorry to report that they just can't be too sure.

The Big Bang Theory

As most of you have probably already figured out, this theory of an expanding and cooling universe is commonly known as the *Big Bang theory*. Ironically, the early proponents of the idea didn't give it that name. That's what the defenders of the static universe labeled it. I guess the mud stuck, because the *Big Bang* is what everybody still calls it today, both its supporters and its opponents. Because of its growing popularity and widespread acceptance, we would be remiss not to talk about it. If you want to have an intelligent conversation with a secular scientist about the cosmos, it might be good to know a little of the lingo.

The Big Bang is often portrayed as a theory about the *beginning* or the *origin* of the universe, but that is really not an accurate definition. In fact, I think Christians should especially try to avoid characterizing the Big Bang this way. The main problem with this description is that it gives the impression that the Big Bang is just another alternative to special creation, which it is not. The Big Bang is a serious attempt to explain the obvious astrophysical data with some kind of coherent story that allows the laws of nature to operate continuously over the entire course of cosmic history.

Trouble begins when the Big Bang is intentionally portrayed as an *atheistic* alternative to creation. This is about as silly as claiming that medicine is an atheistic alternative to healing. Christians don't make these kinds of silly statements about other scientific theories. Take *germ theory* for example. If a doctor were to rely on unexplainable supernatural phenomena to heal an infection, he would be considered a "witch doctor" and immediately have his medical license revoked. So just like serious doctors who seek to explain sickness and disease in terms of natural cause and effect, serious cosmologists seek to explain the astrophysical data without supernatural interference.

If we truly believe that God is sovereign over His creation, then Christians should not fear these naturalistic explanations. God is just as present through the continuous operation of the laws of nature as He is with miracles. And if the story of cosmic history does include any miracles, then the discernable patterns of providence that we call physics might even reveal some obvious discontinuities that defy a cause-and-effect explanation. But even then, honest science demands that we continue the search for a material explanation until one is found. So when Christians accuse cosmologists of intentionally leaving God out of the picture, it does not encourage constructive dialogue between science and religion. We can avoid this by simply adjusting our theological expectations before asking science to describe the formational history of the cosmos. Quite simply, if you ask a scientific question, you should expect to get a scientific answer.

The result of this misunderstanding is that many Christians feel compelled to dispute the Big Bang on *religious* grounds as a godless version of creation. Again, this makes about as much sense as rejecting meteorology as a godless view of the weather. The Big Bang is just an idea that tries to explain what scientists observe in nature with a useful physical model that relates observed *effects* with material *causes*—as far back as physics can reasonably reach. If anybody thinks they have a better way to explain the changing structure and properties of the cosmos over time, then they are free to challenge the Big Bang theory on *scientific* grounds. In fact, there have been a few opposing theories that have made their way into the professional journals. But attacking the *science* of the Big Bang on *religious* grounds is extremely counterproductive.

I completely understand that the motivation is simply to "stick up" for the Bible over any ideas that seem to leave God out of creation, but consider this: when Moses gave Genesis to the ancient Hebrews, was the focus of the biblical creation narrative to combat their erroneous *cosmology* or the pagan *theology* of their polytheistic oppressors? Was God really that concerned about the Hebrew view of a flat earth, a solid firmament, the waters above the heavens and a geocentric universe? Or was His emphasis on building a proper theology of creation by correcting the polytheistic rendering of the ancient Near-Eastern cosmos and establishing Yahweh as the creator and sustainer of all things?

If God, who has perfect knowledge of the universe He created, did not elect to challenge the ancient Near-Eastern cosmology, why then do modern Christians spend so much time and energy combating the contemporary scientific renderings of the natural world when the real problem today is no different than it was back then? Just like God's people of old who adopted the worldview of their pagan neighbors, modern Christians have embraced the pagan philosophies of the world. Of course, we are much more sophisticated today than the ancients were. We prefer a deistic or atheistic creation theology since polytheism was so "yesterday."

Because science is not really concerned with metaphysical questions, like why there is *something* instead of *nothing* or how *something* could have come from *nothing*, the Big Bang theory is not too concerned about the *origin* of the universe, just its *evolution* over time. Anytime you hear astrophysicists talk about *origins*, they are most likely referring to the origin of matter and structure from within the changing universe, not the origin of the universe itself. This isn't just semantics either. There is a big difference between the two. The reason for this is simple. Science is helpless without laws, and since there are no laws of physics that can reach back to *t=0* (the very instant of the Big Bang event), it would be pretty difficult for science to speculate on *how* or *why* the universe came into existence.

You see, the Big Bang event was not just the instant at which all of matter came into existence out of nothing; it was also the beginning of *time* itself, along with *space*. So it does science no good to talk about what happened before the Big Bang. The use of the term *before* necessarily implies the continuity of time between cause and effect, and there wasn't any time until after *t=0*. It also makes no sense to talk about the *cause* of the Big Bang because *cause* also implies *before*, which is impossible without time. You can't have cause and effect with no time or space for a cause to operate. As far as science is concerned, the universe started with an *effect*. So it's not like the universe has or doesn't have a cause. It's more like we're wrong just to ask the question. In fact, as soon as we start thinking about these things, we're probably wrong. So forget about it. Our brains just can't handle it. The important thing to remember is that the Big Bang is simply a theory about how the universe changes, or evolves, over time. A more accurate (and unfortunately more offensive) term is *cosmic evolution*.

These common misconceptions about the Big Bang contribute to at least some of the negative reactions by people of faith. The theory is often perceived to be in competition with divine creation, but it actually makes no such claims of ultimate origins. The chain of natural cause and effect breaks down at the beginning of time. If somebody wants to speculate on the cause of the Big Bang, they have to use philosophy or theology. These tools of reason transcend ordinary time and are not subject to physics. The only consequence to doing this is that any answer provided by philosophy or theology is not *scientific*, but *religious*. And that's perfectly ok. Not every truth has to be a scientific fact. We know plenty of things that have no material explanation. Since any ideas about the *cause* of the universe can't be falsified or established by science, they must be accepted by *faith*.

A Flash from the Past

In order for the Big Bang theory to gain widespread acceptance among the naturally skeptical scientific community, it needed more supporting evidence than just *galactic redshift* and some *equations of state*. So several scientists put their heads together to try to find some other clues to the universe's tiny, hot beginnings. The result was the development of a testable prediction.

If our universe really is expanding and cooling, then a very significant event would have taken place when its temperature dropped below about 3,000 degrees.[16] Before that point, the chaotic soup of protons, neutrons and electrons that are thought to have made up the infant cosmos still had too much energy to form stable atoms. So these highly energized particles flew around the universe, smashing into one another and scattering photons (light particles) in every direction giving space a characteristic glow.

This pervasive glowing plasma also acted like a kind of luminous fog, preventing light to travel freely. Imagine yourself inside of big furnace engulfed by a giant ball of fire. I know it hurts, but bear with me here. In this furnace, you couldn't see your hand in front of your face because the space between your hand and face just glows hot like the sun. If the Big Bang actually happened, that's what the universe would have looked like just before it cooled below 3,000 degrees.

Knowing this doesn't require any speculation or imagination. A temperature of 3,000 degrees is not something ridiculously hot that requires special theories and assumptions to understand. The behavior of ordinary matter under these conditions can be observed in a laboratory with good precision. Because it is so well understood, in 1948 a physicist named George Gamow predicted that we should still be able to detect the 3,000 degree glow of the early universe. As the temperature of the early universe fell below 3,000 degrees, the individual particles that make up ordinary matter would have settled down enough to form stable atoms.[17] At that moment, according to the prediction, everything stopped glowing, the primordial "fog" lifted, and the universe became transparent to light. And those last photons of light from the early glowing universe could have then freely traveled across the cosmos in every direction.

The prediction was made that if the Big Bang really happened, this "snapshot" of the early universe should still be flying around the cosmos today. This was a perfect test for the new theory. What made it so perfect is that this "baby picture" of the cosmos would have to exhibit very specific properties, not just any "flash of light" would do.[18] Two other physicists, Ralph Alpher and Robert

---

[16] Physicists like to measure temperature in *Kelvin* rather than degrees. The Kelvin scale is easier to work with than degrees because it starts at absolute zero. This avoids having negative temperatures. On the Kelvin scale, zero is the absolute lowest that you can go. It's about 273 degrees below zero on the Celsius scale. However, since I don't want to confuse our non-scientific friends, I'll just say *degrees*.

[17] The first atoms to form in the early universe were the two simplest: hydrogen (90%) and helium (10%).

[18] For you advanced readers, the CMBR would have to exhibit "black body" characteristics of a 3000K source redshifted to the microwave range (peak frequency), be uniform in every direction, and be free from any spectral absorption lines that might result from passing through interstellar clouds of dust and gas—which didn't exist until after the universe became transparent.

Herman actually calculated what this radiation would look like today. If the new theory about the expanding cosmos was true, then the light waves from this 3,000 degree glow would be stretched out quite a bit by now. In fact, they calculated that it would look more like a 5 degree glow today because its radiation would be stretched out past the red end of the spectrum, past the infrared region, right smack into microwave territory. And there are no other known cosmic phenomena that could produce a similar pattern of microwave radiation.

Scientific showdowns don't get any better than this! These three physicists basically threw down the cosmic gauntlet. If the Big Bang actually happened, the universe should be uniformly bathed in a very specific pattern of microwave background radiation. Such a discovery would have no other physical explanation but an expanding and cooling universe that was once very hot and small.

This is a good opportunity to bring up an important point about the nature of scientific evidence. You may often hear people ridicule theories like the Big Bang because they can't be directly observed. In debates about the Big Bang, you may even see creationists make statements like "well, were you there to see it?" This might score a few points with people who are less informed, but it is basically no different than our hypothetical neighbor from Chapter One who claimed that there can be no way to prove that it snowed last night unless somebody was awake to see it actually falling from the sky—despite the indirect evidence of 18 inches of white powder. Unfortunately, these tactics are often effective on people who are looking for reasons to ignore scientific claims. Once the jury has been thoroughly confused by this "reasonable doubt" approach, the defense can then offer its own version of events, like our neighbor and his hypothetical fleet of levitating snow machines.

In the real world, scientists are rarely afforded the luxury of direct observation. Most things have to be inferred by what few things can be observed. In the case of something that happened in the distant past and can not be repeated, a valid theory must be able to make falsifiable predictions about what we would expect to see if the theory were true. In other words, if the Big Bang theory is true, then we should find X or Y today. These kinds of tests can provide strong evidence in favor of the claim, especially when X or Y can have no other physical explanation. This would be similar to making the prediction, before going outside, that if it really did snow last night, I should find snow still on the ground since there is no other known natural phenomenon that causes the effect of snow (levitating snow machines notwithstanding). In doing this, you can test your theory by just walking outside and making an indirect observation, even though your opportunity to directly observe the event while it's happening no longer exists.

So did the Big Bang theory pass the astronomers' test? Was the predicted radiation ever detected? In 1964, two physicists named Arno Penzias and Robert Wilson were working for Bell Telephone Laboratories on communications satellites. These early satellites used microwave radiation to send and receive information from earth-based facilities. One problem that they kept running in to was an excessive amount of static that appeared to be coming from empty space. So they built an enormous receiver that could detect even the faintest microwave signals in an attempt to gather some data on this annoying static. No matter where above the horizon they pointed their antenna, they measured the same static. Ironically, these guys were not cosmologists and had no clue about the Big Bang predictions of the Cosmic Microwave Background Radiation (CMBR) made nearly sixteen years before.

They published a paper in 1965 complaining about microwave static, but little did they know that what they stumbled upon was the greatest astronomical discovery of the century! The discovery of the CMBR changed cosmology from a theoretical science to a hard science with analytical data. The elusive snapshot of the early universe was finally discovered! The information provided by the CMBR revolutionized Big Bang cosmology. Astronomy would never be the same.

There were still many astronomers who were holding on to the static theory of the universe and were very skeptical of the newly discovered CMBR. So by 1965 physicists and astronomers everywhere had turned their attention to studying this phenomenon. The current temperature was calculated to be 2.73 degrees, slightly off from the temperature predicted in 1948, but entirely consistent with the Big Bang predictions of the 1960s. In fact, the CMBR passed every test it was given. It had the right temperature, it was uniform in every direction, it was independent of the earth's motion, and had the exact distribution of microwave frequencies predicted by the theory. By the end of the 1960s, the Big Bang theory had endured 40 years of scientific scrutiny, fulfilled every cosmological prediction, and ultimately became the mainstream scientific view of cosmic evolution.

The accidental discovery of the CMBR earned Penzias and Wilson the Nobel Prize for Physics in 1978. With this data in hand, astrophysicists were able to accurately determine several other cosmic parameters that were previously unknown. Over time, many scientists have used these parameters to make more and more accurate predictions about the changing rate of cosmic expansion. The current estimate of a 13.7 billion year old universe is a direct result of this ongoing process.[19]

---

[19] If you want the full story in exhaustive detail, I recommend the book, *Origins: Fourteen Billion Years of Cosmic Evolution* by Neil deGrasse Tyson and Donald Goldsmith, (New York, NY; Norton, 2004).

Other Appearances of Cosmic Age

For many scientists, the evidence for an old universe doesn't end with the Big Bang theory. There are many other simple observations that confirm the antiquity of the cosmos with great precision. In fact, modern astronomy allows direct observation of just about any moment of cosmic history, except for the time before stars were first formed.[20] How is this possible? How can astronomers rewind the universe back to any point in time they want and watch things unfold? Is there some kind of magic time machine that allows people to look into the distant past?

Why, yes, in fact there is such a device, but it's not magic. It's called a telescope. It is a very simple "time machine" based on two parameters: light and distance. The finite speed of light, traveling over great distances, takes time to get from one place to another. Therefore, if one wants to look back in time, they only need to look deep into space. Actually, you don't even need a telescope to look back in time. The sun is a little more than eight light-minutes away. When you look at the sun (don't look for too long), you are not seeing as it is now, but as it was eight minutes ago. That's how long it takes the sun's light to reach earth. If our sun suddenly disappeared, nobody would even know it for eight minutes.

Everything we know about physics suggests that the speed of light through space is a constant 186,000 miles per second (more on this later). The speed of light may seem pretty fast to us here on earth, but on cosmological scales, it is actually quite slow. For instance, it takes over 100,000 years for light to cross our own galaxy. Think about that next time you look up at the Milky Way! Our nearest large extragalactic neighbor, the Andromeda Galaxy, is over 2 million light-years away. When we look at Andromeda today, we don't see it as it is now, but as it was over 2 million years ago, and so on and so forth. The deeper we look into space, the further back we look in time. That is our astronomical time machine. As bigger and better telescopes are developed, our ability to directly examine the early universe gets better and better.

So what do astronomers see when they look back in time? They see more evidence of a very old universe with a long and fascinating natural history. In our own galaxy, new stars being born from clouds of interstellar dust and gas can be observed. Millions of "middle aged" stars, like our own sun, burning through their nuclear fuel at different rates and temperatures according to their size and mass can also be seen. In fact, astronomers have collected

---

[20] According to the Big Bang theory, the first stars started to form after the first few hundred million years. Before that time, the only thing that was around to create light for us to observe was the CMBR.

enough data on the stars in our own galaxy to accurately characterize the entire stellar lifecycle of any star in the universe from birth, through the main sequence, to death. This process can take anywhere from a few million to a few trillion years to unfold, depending on the size of the star. The first smaller stars that formed can still be seen burning today, while several generations of the larger more massive stars appear to have come and gone in great explosions, distributing their cremated remains across the cosmos and leaving behind beautiful glowing clouds. Clusters of hundreds of thousands of stars can be seen grouped together and each individual star appears to have the same birthday as the others, indicating that they were all formed out of the same cloud of dust and gas.

In far away galaxies, evidence of old stars dying can be seen by observing huge luminous explosions that briefly outshine the entire galaxy they inhabit. The most distant one observed to date is estimated to be 10 billion light-years away.[21] The fact that an object 10 billion light-years away can even be observed suggests that the light reaching our eyes has been traveling through space for 10 billion years. Astronomers have also seen fully formed galaxies in the process of colliding with each other, appearing to be frozen in time by their enormous size but literally morphing into a single galactic system. The fact that events can be observed that required billions of years to unfold testifies to the apparent long, rich, natural history of the cosmos.

Observing cosmic history by looking deep into space has also provided astronomers with visual evidence that confirms many of their previous theoretical predictions. The older-looking nearby galaxies are seen as they were only a few million years ago, having fully evolved spiral structures like our own Milky Way. Conversely, the younger-looking far away galaxies are seen as they were billions of years ago having only simple, unorganized, and nebulous structures, entirely consistent with what one would expect to find in a younger, less developed universe.[22] The list goes on and on. No matter how scientists look at the evidence, the universe appears to have existed, not for an eternity, and not for only a few thousand years, but for tens of billions of years, 13.7 to be precise.

---

[21] Supernova 1997ff

[22] Recall our discussion from Chapter One. Don't be misled by incorrect applications of the 2nd law of thermodynamics. Under the right conditions, organization and structure can emerge from chaos. Next time a mature, well developed hurricane hits the coast, look at how organized its structures are, like the characteristic "eye" of the storm. Compare this "highly evolved state" with the complete disorder of a tropical depression, which resembles nothing more than a thunderstorm. These structures "evolve" from a tropical depression, to a tropical storm, and finally to a hurricane. If other forms of natural evolution could be as easily observed, there probably wouldn't be such a fuss over it.

Thousands or Billions?

So how do we square all of this with the biblical data? Is there a scientific "smoking gun" that can bring down this cosmological house of cards? What is the scientific evidence for a 6,000 year-old universe that exposes modern science as a fraud and confirms the testimony of Scripture? These are some of the serious questions asked by many Christians when hearing the details of cosmic antiquity and its evolution over time. I've certainly asked them. In fact, these were some of my motivations for studying the Big Bang theory in the first place. But once we start asking these kinds of questions, I think we've already missed the point. Actually, we've missed two important points: the point of the Genesis creation account and the point of the Big Bang theory.

In Part II of the book, we looked at the point of Genesis as giving us a biblical theology of creation. So what then is the point of the Big Bang? Isn't it just another way to attack Christianity and the authority of the Bible, or to keep God out of the public schools? For the longest time that's pretty much what I thought. Judging by the rhetoric of most creationist literature, I'd say I wasn't the only one. Of course, any scientific theory can be hijacked by atheistic presuppositions for the sole purpose of arguing against God and creation. This is a common tactic used by the enemies of Religion.[23] But is that really *the point* of the Big Bang theory? Can we be a little more gracious than that? Is it possible that the Big Bang is really nothing more than a tentative natural cause-and-effect explanation of what scientists observe in nature? It seems to me like a pretty coherent way to explain things like cosmic redshift and the CMBR in terms of the known patterns of material behavior that we call the laws of nature. And isn't explaining the discernable patterns of natural behavior *the point* of any scientific discipline, from medicine to meteorology?

Whether we agree with it or not, there are many very intelligent people who consider the Big Bang theory one of the crowning achievements of modern science. It represents decades of hard work, a little good fortune, and an unparalleled cooperation between the world's leading astrophysicists, particle physicists, and nuclear physicists. The reason it has such a broad appeal to the scientific community is not because it makes no reference to God, but because it considers all of the observational and experimental data and attempts to provide a coherent natural explanation that assumes the known laws of physics have been operating continuously since the beginning of time. This is

---

[23] Given the view that biblical theism is the underlying principle of the uniformity of nature, these scientific attacks on religion seem somewhat counterproductive. Like cutting off the branch that supports your tree house!

why the mainstream scientific community can say, with some degree of confidence (which is often perceived as arrogance), that 13.7 billion years ago the universe was very small and very hot.

One thing that Christians need to understand is that these are firmly held beliefs based on what appears to be overwhelming physical evidence. The Big Bang is not just something that atheists can conveniently substitute for special creation to avoid the existence of God any more than gravity is something they can exchange for providence to avoid God's sovereignty. Believe it or not, scientific evidence has little to do with how most people feel about doctrines like creation and providence. Is that the reason why you believe? Did you carefully weigh all of the scientific evidence for and against these precious doctrines before coming to the conclusion that you must put your faith in Christ? Or did you accept these things *by faith* on the testimony of Holy Scripture? So when a very intelligent scientist rejects God and the Bible in favor of materialism or atheism, there are usually deeper issues than their specific interpretation of galactic redshift or the CMBR.

Consequently, attacking the Big Bang theory as a "silly creation myth invented by people who hate God" does not address the real problem. These tactics assume that naturalistic theories are incompatible with religious faith and are themselves the root of the problem. I submit to you that they are not. Christians don't take this approach to other areas of science. Do our naturalistic theories of biological development from a single fertilized cell to a fully grown adult undermine our belief that God made us? Do our naturalistic theories of planetary motion or the water cycle undermine our belief in common grace?[24] Do our naturalistic theories of germs and antibiotics undermine the need to ask God for healing?[25]

---

[24] Matthew 5:45: "...He [God] causes his sun to rise on the evil and the good, and *sends rain* on the righteous and the unrighteous" (emphasis mine).

[25] Psalm 103:3

CHAPTER SEVEN

THE CHALLENGE OF COSMIC HISTORY

For those of us that have been raised to believe that the entire cosmos is only a few thousand years old (and I include myself in this category), the scientific evidence presents a challenging problem. If we take the popular approach and assume that Genesis is giving us creation science, then we must forge some sort of agreement between science and the Bible. In order to accomplish this difficult synthesis, Christians will either have to find a way to explain away the apparent 13.7 billion-year natural history of the cosmos, or we will have to find a way to read the Bible so that it better supports the Big Bang theory. If we can't pull this off, the world will try to use this evidence to dismiss God's Word as false and unreliable. The stakes are high!

This is a tall order for the creation scientists, and whether they succeed or fail, the consequences are not good for the church. In addition to isolating Christianity from mainstream science, creation science also divides Christians into two camps that spend as much time fighting each other as they do fighting the evolutionists. These two opposing sides of creation science are called the Young-Earth Creationists (YECs) and the Old-Earth Creationists (OECs).

Counting Rings on Eden's Trees

The YECs favor the biblical data over the astrophysical data and unapologetically conclude that the universe is six to twelve thousand years old. As a result, they must explain away all physical evidence to the contrary. One simple way to do this is to dismiss all evidence of cosmic history as an illusion. This line of reasoning is sometimes called the "appearance of age" argument. Logically, the "appearance of age" argument makes perfect sense. Think about the Garden of Eden the day after God created the world. The plants, animals and stones all appeared as if they had existed for ages. If Adam had cut down the largest tree in the Garden and counted its rings, there would have been unmistakable "evidence" that the tree existed for many years, even though it was actually only a few days old. So any evidence found for an old earth is expected since a coherent natural history would have accompanied anything that was created in its present form. Is our problem solved? Perhaps it is for some.

The real problem with this argument is not logical, but *theological*. Not only did God have to create trees with rings in the garden, but he apparently had to create 10 to 12 billion years of electromagnetic radiation (light) streaming in from all parts of the distant cosmos. How else could one explain the appearance of stars and galaxies further than 6,000 light-years away? This imaginary light would have contained a fascinating history of events, such as the CMBR, that never actually transpired, but were specially fabricated by God to deceive us. This goes beyond just counting the rings on Eden's trees. This would be like God implanting false memories of childhood events that never happened in the minds of Adam and Eve.

All of this is certainly possible with miracles (because anything is possible with miracles), but none of it paints a theologically acceptable picture of God as the Creator of grand illusions designed to intentionally mislead people. Moreover, if all things were created with a coherent natural history, then the only way to tell *apparent* history from *authentic* history is by consulting the Scriptures. Anything that happened prior to the creation week would be considered *apparent* natural history, and anything taking place after the creation week would be considered *authentic* natural history.

This brings up an interesting situation: if only the Bible can decide which scientific observations about the earth are *authentic* and which are *apparent*, then perhaps the 67 geocentric verses of the Bible should not be made to conform to scientific observation either? Maybe the earth only "appears" to revolve around the sun and rotate on its axis when it actually "can not be moved" as the Scriptures teach? This might seem silly to you, but the "appearance of heliocentricity" argument is actually used by the *Association of Biblical Astronomy* in support of a geocentric universe. Or perhaps the moon only "appears" to reflect light from the sun when it actually is a great light as Moses describes it. You can see how ridiculous this approach can be.

Another way for YECs to challenge the evidence for cosmic history is to protest the speed of light as a universal constant by claiming that is was once much faster in the past. If Christians can prove this, then they can explain how all of the light from these distant objects has reached us in just a few thousand years. Arguments along these lines are very common among Young-Earth Creationists and are sometimes called "speed of light decay" theories. Certainly there are a few smart physicists with PhDs working hard to solve this "problem," but there are also a few smart astronomers with PhDs who are still trying to show that the entire universe revolves around the earth based on a few verses of Scripture. So beware of eccentric scientists with PhDs making outrageous claims based on questionable interpretations of the Bible.

Whenever you see arguments against the constancy of the speed of light, just remember this: the speed of light through the vacuum of empty space is a fundamental constant of nature. Anybody who wishes to challenge it is going to need some pretty solid evidence in order to be taken seriously. And if the speed of light was much faster in the past, there should be some evidence of it.

Modern technology enables physicists to measure the speed of light with incredible precision. Even the slightest variation could be detected. Since about 1973, measurements of the speed of light using lasers have always come out the same.[1] So if the speed of light were much faster in the past, it must have decayed rather quickly in order to level off at its current value. Obviously if it were still decaying, our current measurement techniques are good enough to detect it. So any theory about the changing speed of light needs to account for both its previous *decay* and its current *stability*.[2]

In order for us to see objects that are 10 billion light-years away in only 6,000 years, like Supernova 1997ff, the speed of light would have needed to start out extremely high. In nature, these kinds of quickly falling quantities that naturally level off are described by something called *exponential decay* and are actually quite common.[3] If this is what has indeed happened to the speed of light, it's not that hard to figure out what the decay rate and the initial velocity needed to be. After a little math, we find that when God spoke the words, "Let there be light," that very light would have been almost 29 million times faster than the light we see today![4]

---

[1] Before the use of lasers, precise measurements of the speed of light were difficult and varied because of experimental error. Some proponents of "speed of light decay" have hand selected a few of these higher readings to show that the speed of light was once faster than it is today. After ten years of consistently measuring the speed of light to be unchanged, in 1983 it became the standard by which the meter is determined. Consequently, it has pretty much been a fixed value since then and is no longer measured.

[2] Since atomic clocks are also used to measure the speed of light, some critics claim that this set-up uses "light" to measure "light" because the rate of nuclear decay also paces itself from the speed of light. This is a good point, but if there were any evidence that the speed of light was presently decaying, it would make headlines all over the place.

[3] One of the problems with the exponential decay rates found in nature is that they usually level off at zero as time goes to infinity. So the speed of light would still be slightly decaying if truly followed the rules of exponential decay, which doesn't appear to be the case. To get around this obvious problem, proponents of the speed of light decay theory created "custom" decay rates like $\log(c)=A+B[\log\sin(t)]$, which give better results, but do not resemble anything else found in nature. Pulling these things out of thin air without precedent is enough to make most scientists nervous about these kinds of theories.

[4] For scientifically advanced readers who want to play around with the decay equations:
- The velocity of light at any year *(t)* since creation: $V(t) = V_0 e^{-Kt}$
- The distance in lightyears that the first beams of light traveled for any year *(t)* since creation:
  $S(t) = 10^{10}(1 - e^{-Kt})$
- $V_0 = 28,615,783c$ (The initial velocity for light)
- $K = 0.00286158$ (the decay rate parameter)

These equations are fitted to accommodate what is already known about the speed of light. For instance, $S(t) = 10^{10}$ at $t=6,000$ (1980) so by 1980 we can see objects that are 10 billion light-years away after only 6,000 years. Also the speed of light, $V(t)$, reaches is present value of c by 1980 $(t=6,000)$. These equations assume a creation date of 4020 B.C. so the universe was 6,000 years old in 1980, but you can adjust it for any date of creation you want.

Problem solved! Or is it? Sometimes by solving one problem, you can un-intentionally create about a dozen others. Unfortunately, a speed of light this fast would completely wreck the coherence of general and special relativity. The delicate relationship between mass, light speed and energy can not be easily upset. For starters, Adam and Eve would have been fried to death by radiation from the incredibly high rates of nuclear decay that pace themselves using the speed of light. While this might account for the observed quantities of certain isotopes present in the earth's crust in relation to a young earth, it doesn't explain how the Garden of Eden could have survived a nuclear melt-down.

So in order to accept this theory of speed of light, or *c*-decay, you also have to assume that God suspended the other known laws of physics just to bring us light from the far reaches of the newly created cosmos. And if you're going to toss out the continuous operation of the laws of nature and replace them with miracles, then what's the point of trying to come up with a scientific theory, like *c*-decay, that attempts to explain it all in the first place? I've never understood this approach. In order to solve one problem using a consistent scientific methodology like exponential *c*-decay, the YECs will create a dozen other problems that literally need miracles to fix. Why not just invoke a miracle to solve your first problem and spare the rest of the universe from gross physical incoherence?

Miracles don't follow the rules of science and don't need scientific proof. If the YECs are going to ultimately sprinkle magic fairy dust on the whole problem to make it work, then they should forget about trying to prove these things with science and just use the "appearance of age" argument. It also relies on a miracle (a miracle of deception) to overcome the problem of distant starlight, but without compromising the intelligibility of the universe at the same time. If Occam's razor were to allow for miracles, it would prefer just one that takes care of everything, rather than many independent miracles all trying to accomplish the same objective. The "appearance of age" miracle allows Adam and Eve to see everything we see, without having to wait thousands of years for the light to show up, and without being toasted from the intense radiation caused by a 30 million-fold increase in nuclear decay rates.

Nevertheless, if anyone can demonstrate, with a sound theoretical model backed up by observational data that the speed of light has been rapidly decreasing over the past 6,000 years, fame and fortune (and definitely a Nobel Prize) awaits them. Whoever can prove this would literally be the *Einstein* of the 21st century! Any takers? The fact that this hasn't happened yet has nothing to do with secular "conspiracy theories" against legitimate science that supports biblical claims. Believe it or not, most scientists really don't care about that. If the theory can make testable predictions, then all they want to know is whether or not the data support the predictions. That's it.

The Big Bang is a perfect example of this. When the idea was first proposed that the universe had a beginning, some in the scientific community feared that it would be a kind of "back door" for biblical creationism. For a while, scientific theories about the *beginning* of the universe were referred to as the "Genesis problem." Others called them "philosophically unacceptable." It seems silly today to think that some scientists once feared the Big Bang because it agreed with the Judeo-Christian idea of creation from nothing, but this shouldn't come as a surprise. New scientific theories are always initially opposed by the defenders of the old system, sometimes on religious or philosophical grounds, but always on scientific grounds. After all, those that worked so hard on the passing theory aren't just going down without a fight!

As it turns out, all of the predictions made by "speed of light decay" theories have been directly contradicted by the data over and over again. That is why c-decay theories are not taken seriously by professional scientists, even the majority of professional Christian scientists—not because of a conspiracy against any evidence that supports biblical creation, but because there is no evidence that shows the speed of light has decreased from an extremely high value. That's just how science works. If the observations support the theory, then the theory survives another day. If the observations can neither confirm nor deny the theory, then we keep it at arms length until it can be substantiated. If the observations directly contradict the theory, then we discard it.

So if a genius like Albert Einstein hesitates to come forward and be the first to predict an expanding universe because of a lack of data, even though his original cosmological equations described an expanding universe, then how is it that some YECs have the audacity to assert bold claims about fundamental constants of nature despite the solid evidence *against* them? Again, bold claims that can potentially overturn years of basic scientific understanding usually require more than a theoretical prediction; they require hard data, and lots of it.

I'm convinced that the only reason this theory about c-decay is still hanging around is because of the misguided philosophical and religious motivations of creation science, not because it has been confirmed by any evidence. I'm also convinced that this kind of behavior among YEC scientists is a serious stumbling block to the Gospel message, especially among scientists who take their work seriously. When the YECs habitually disregard the basic protocols of scientific scholarship, they are basically telling the world that Christians don't need to play by the rules. They don't need go through the peer-review process or present their theories at international conferences where other world renowned experts in these fields can evaluate them. They don't even need data or evidence to support their theories because *God is on their side*!

Remember that all scientific conclusions are tentative. Scientific theories and hypotheses are always subject to revision or replacement, but there is nothing tentative about creation science. Any evidence to the contrary is ignored without consideration. I can honestly say that I see more humility when I read secular scientific literature than when I read YEC literature and it's embarrassing. Is this the mark of a good missionary? Aren't Christians supposed to respect the customs and traditions of the culture that they are trying to reach? Will scientists ever take seriously our claims about the person and work of Jesus Christ if some of us keep pushing these questionable scientific theories through the back door? At some point, after these same arguments keep falling on deaf ears, the YECs need to step back, regroup, and refocus on what's really important, on what the point of all this really is.

Squeezing Scientific Blood from the Biblical Turnip

Old-Earth Creationism is the other side to the Creation Science coin. It embraces the obvious antiquity of the cosmos without any scientific reservations, but insists that there is no contradiction with the Bible. In fact, some even claim that the Bible actually teaches the Big Bang theory. Apparently, it has been right there under out noses all of these years! In order to demonstrate this, a substantial amount of Scripture has to be taken out of its original context in order to support modern cosmology, setting a dangerous hermeneutical precedent.

Rather than just leave those primitive descriptions in their ancient Near-Eastern context, great efforts are made to read modern scientific concepts between the lines of the Bible. For instance, the many verses in the Psalms[5], Isaiah[6], Job[7], Jeremiah[8], and Zechariah[9] that refer to God "stretching out the heavens" are interpreted by some to describe cosmic expansion and Big Bang theory. Great hermeneutical care is taken to ensure that the subtle nuances of the Hebrew word for "stretch" proves conclusively that cosmic expansion is not finished, but is ongoing.[10] I'm sure this news comes as a big relief to those whose faith was hanging on whether or not the cosmic expansion taught by the Bible was in agreement with the latest CMBR data from the Wilkinson Microwave Anisotropy Probe.

---

[5] Psalm 104:2

[6] Isaiah 40:22; 42:5; 44:24; 45:12; 48:13; 51:13

[7] Job 9:8

[8] Jeremiah 10:12; 51:15

[9] Zechariah 12:1

[10] Hugh Ross, *The Creator and the Cosmos* (Colorado Springs, CO; Navpress, 1993), pp 24-25. This same argument can be found at http://www.reasons.org/resources/fff/2000issue03/index. shtml#big_bang_the_bible_taught_it_first

Some Christians may not be aware that the Apostle Paul was also quite the astrophysicist. In his epistle to the Romans, Paul apparently makes the first case for the cosmic background radiation based on the concept of an expanding universe taught by the Old Testament Prophets. Consider the following:

> Finally, the Bible indirectly argues for a big bang universe by stating that the laws of thermodynamics, gravity, and electromagnetism have universally operated throughout the universe since the cosmic creation event itself. In Romans 8 we are told that the entire creation has been subjected to the law of decay (the second law of thermodynamics). This law in the context of an expanding universe establishes that the cosmos was much hotter in the past.[11]

Does anybody seriously think that these ancient writers were trying to tell us that the cosmos was once very tiny and hot? If so, then why didn't we figure this out before the astronomers? Shouldn't at least one of those monks who spent their lives copying the Bible by hand have figured this out first? Why hasn't the Big Bang theory been a part of Christian doctrine from the beginning?

I know that this kind of stuff is fun and can provide hours of entertainment, like trying to figure out which world leader is going to be the antichrist, but trying to read modern cosmology into these verses completely misses the point. By ripping these verses out of their original ancient context, we inadvertently place them on the altar of scientific naturalism only to be picked apart by the critics of the Bible. What if a future discovery were to turn the Big Bang theory upside down? Where then would that leave us? Remember: all of science is tentative, but the Word of the Lord standeth fast forever! Let's not confuse the two.

Like all passages of Scripture that appear to teach creation science, the verses that paint a picture of God "stretching out the heavens" are best understood in their original context. In ancient times, when somebody would pitch a tent, they would "stretch out" the canvas across some beams in the shape of a dome or a vault, which is exactly how the ancients viewed the sky, or the "heavens." Since the heavens were thought to be "God's dwelling place" in the ancient cosmogony, it should come as no surprise that the creation of the biblical firmament, which had nothing to do with outer space as we know it, is described by the human authors as God "stretching out" a tent in which to dwell.

---

[11] Ibid, pg. 26.

Now ask yourself this: what would the ancient nomadic authors of Scripture have been more familiar with, primitive tent construction or cosmic expansion? Psalm 19:4 even says that God "has pitched a tent for the sun," completely consistent with the ancient cosmogony. Psalm 104:2-3 specifically says that God "stretches out the heavens like a tent and lays the beams of his upper chambers on their waters." Again I ask the question, is the author drawing on his own experience with tents, or is he receiving a cosmological "word" from God? Do these "waters above the heavens" belong to modern cosmology or do they belong to the ancient Near-Eastern cosmology? Which one of these interpretations seems more natural and which one seems more contrived?

The Rest of the Story

Is there another way that Christians can respond to the physical data? Perhaps one that doesn't ignore the solid evidence for an old universe as the YEC approach does, but doesn't offer up the Scriptures on the altars of science as the OEC approach seems to do? I think that there is, but before we get to it, let's look at the rest of cosmic history.

If the universe as we know it really did begin as a tiny hot soup of subatomic chaos some 13.7 billion years ago, then it must have gone through an incredible series of transitions to look like it does today. How did everything achieve its present state from those tiny hot beginnings? Did any of these scientists ever stop and think about that? How does a big explosion of heat and light give rise to all of the different elements like carbon, oxygen, nitrogen, and hydrogen? How do complicated structures like galaxies and solar systems emerge from nothing? Can tornadoes blowing through junkyards assemble 747s?[12] Believe it or not, these questions have some very thoughtful answers.

In the Beginning

In the beginning, when the universe was thought to be more than 10^{30} degrees and less than 10^{-45} seconds old, all theories about the behavior of matter are useless. The laws of nature as we know them don't even apply before this time. But somehow the four fundamental forces of nature would have been unified into a single mega-force. Unfortunately, the best modern science

---

[12] This is a popular reference from Fred Hoyle, *The Intelligent Universe* (New York, NY; Holt, Rinehart and Winston, 1983), pp. 18-19. It's a good question, but the problem is that 747s aren't found in nature. So they are obviously not *natural* like tornadoes are. But a tornado is a pretty complicated structure that came from "nothing." So are tornadoes *designed* or do they evolve by the continuous operation of the laws of nature?

can do is combine the electromagnetic and weak nuclear forces. Beyond that, there are still a few Nobel Prizes to be handed out.

The four fundamental forces of nature are gravity, strong nuclear, weak nuclear, and electromagnetic. By these four forces, God governs the entire material universe, from the smallest particle to the largest galactic supercluster. All patterns of material behavior that we observe and describe are a result of the physical interactions between these four fundamental forces. Everybody knows what gravity is and you should also be familiar with electromagnetic forces (electric and magnetic fields), but most people can live happy and fulfilling lives without ever having heard about the other two. That's because the two nuclear forces are only effective at extremely small scales, inside of individual atoms.

The Epoch of Inflation

The first individual force thought to emerge from the unified force was gravity, quickly followed by the strong nuclear force.[13] That event, at about 10^{-35} seconds after the moment of creation, coincided with a 5 trillion percent increase in the size of the universe, smoothing out the energy density to less than one part in a hundred thousand. While this growth spurt would have only made the universe about the size of a softball, the significance of this period of cosmic "inflation" can't be overstated. The distribution of energy over the infant cosmos was the medium from which all matter would eventually emerge. If this distribution were totally smooth, the universe today would just be a thin cloud of hydrogen and helium gas, without any stars or galaxies. If the distribution of energy was just a tad less smooth, then matter would have clumped together too quickly under the influence of gravity, forming only giant black holes. Either way, life as we know it wouldn't exist were it not for something that happened 13.7 billion years ago in less than a blink of an eye.

The Creation of Matter

You've probably seen the famous equation $E=mc^2$ tossed around by smart people (I'm not implying anything about the author). That equation tells us that energy can be converted into matter and that matter can be converted back into energy. Converting matter into energy should be familiar to you. For example, a nuclear power plant *slowly* converts matter into en-

---

[13] The strong nuclear force binds together protons and neutrons in atomic nuclei, overcoming their electromagnetic repulsion.

ergy, while a nuclear weapon *quickly* converts matter into energy. Converting energy back into matter is a little trickier. In our ordinary everyday lives, we don't see matter spontaneously arising from empty space just from the addition of energy. So to many people, the creation of matter from energy might seem like science fiction. However, if you were to raise the thermostat to one trillion degrees, there would be sufficient energy to make pairs of particles and antiparticles from "nothing." And that is exactly what we see from particle accelerators when physicists smash particles into one another generating tremendous energies. The stuff made in the laboratory doesn't hang around very long. These pairs quickly annihilate each other, converting their mass back into energy according to $E=mc^2$, but they last long enough to be observed and that's all that a particle physicist needs to prove the point.

After the inflation event, empty space still contained enough energy to create particle/antiparticle pairs. However, for reasons still unknown to science, there was a small accounting error. For every one billion particle/antiparticle pairs that were created, there was an extra particle with no antiparticle; a *yen* without a *yang*. This slight imbalance in particle production could have had something to do with the next significant event, which was the separation of the electromagnetic and weak nuclear forces,[14] but nobody knows for sure. Regardless of exactly how it happened, after the first 10^{-11} seconds the universe finally had the four fundamental forces of nature as we know them.[15]

As time moved on, the universe kept on expanding and cooling. When the temperature fell below a billion degrees, the spontaneous creation of matter and antimatter ceased.[16] Basically the music stopped and everybody scrambled to find a chair (please excuse the anthropomorphizing, but I'm trying to make this interesting!). As the first second of time passed, all of the remaining particle/antiparticle pairs quickly annihilated each other just like they do in the laboratory. All that remained was the one lone survivor for every billion pairs that came and went. These extra particles would eventually make up all of the ordinary matter in the universe today. Were it not for the apparent billion-to-one accounting error in particle/antiparticle creation and annihilation, the universe would contain nothing but energy today. Solid matter as we know it would have never come into existence.

Was this event a miracle? That is a legitimate philosophical question.

---

[14] The weak nuclear force controls radioactive decay of atomic nuclei.

[15] Everything after 10^{-11} seconds is based on laboratory confirmed physics.

[16] Spontaneous quark/anti-quark production stopped below a trillion degrees, spontaneous electron/positron production stopped below a billion degrees. Quarks would eventually form the protons and neutrons of atomic nuclei.

There appears to be no sufficient scientific explanation for the slight production of matter over antimatter. It is entirely possible that God briefly suspended the laws of nature, leaving behind just enough matter to make the universe interesting. If this was accomplished by a miracle, science would never be able to prove it. The question would remain a scientific mystery forever. But do we gain anything by assuming that this was a miracle? Do we risk anything? Let's think about this for a minute. Whether this billion-to-one accounting error is ever explained by physics or is thought to be a miracle, neither scenario diminishes God's sovereignty over the entire process. So what do Christians stand to gain by attributing it to a miracle?

On the other hand, what do we risk by making these supernatural claims? We know that God is certainly capable of miracles, but for what purpose does He perform them? Does God need to perform miracles to make up for the inability of His creation to operate coherently, or does He perform miracles to get the attention of His creatures and bring glory to Himself? If the latter is true, then who was the audience? Was it 21st century physicists some 14 billion years later? If the former is true, then what does that imply about God's omniscience and omnipotence? Moreover, even if we claim this event occurred by a miracle, the scientific search for a natural explanation will continue. What happens when one is found?[17] Will people then dismiss God as unnecessary? Sometimes I think Christians get caught up in the moment and don't put these kinds of issues into perspective. Again, I'm not saying that it was or wasn't a miracle, I'm just not sure what the advantage is to the "miracle" claim.

As the temperature fell below 100 million degrees, some of the surviving particles joined together to form protons and neutrons, the two basic particles that make up atomic nuclei. For the next two or three minutes, the universe was hot enough to fuse hydrogen nuclei into a single helium nucleus, the next element of the periodic table. There was also just enough energy to fuse some hydrogen and helium into lithium, the third element on the periodic table. There wasn't enough energy to do much more than that.

So after the first few minutes of creation, the periodic table only contained three elements.[18] They were hydrogen (90%), helium (10%) and trace amounts of lithium. That's it. Unfortunately, helium doesn't combine with

---

[17] The small surplus of matter over antimatter that existed after the Big Bang was probably due to the asymmetry of the weak nuclear force (antimatter is not perfectly the mirror image of matter). Unlike the other forces which are unaffected by a reflection in a mirror, the weak force has a definite "handedness."

[18] Though not yet in elemental form because the electrons were still too hot to make stable orbits around the nuclei.

anything and there is not much you can do with mostly hydrogen gas and a few lithium atoms. Without a way to reheat the cosmos and fuse some heavier elements out of these, the universe would be a pretty uninteresting place today.

The Seeds of Cosmic Structure

The next 380,000 years were relatively uneventful in terms of physics, except for the continuous expanding and cooling of the universe of course. But when the temperature dropped below 3,000 degrees the previously glowing universe became a transparent universe.[19] The last photons (particles of light) scattered from that 3,000 degree glow have been bathing the universe ever since. Today, we see this faint redshifted "glow" as the Cosmic Microwave Background Radiation (CMBR).

We talked earlier about the significance of the CMBR as unmistakable evidence of the universe's tiny hot beginnings, but this "snapshot" of the early universe also contains other important information. The CMBR is hard data on how the first atoms in the universe would have been distributed the instant that the universe became transparent to light.[20]

Since the preceding epoch of inflation is believed to have smoothed out the energy from which all matter would eventually arise, any variations in the CMBR were predicted to be extremely small, too small to observe from Earth. But scientists knew that these tiny density fluctuations had to exist or the universe would not have any large scale structure. These first pockets of slightly higher gas density would literally be the seeds from which stars and galaxies would grow. This is another case of the Big Bang theory making a testable prediction. If the CMBR were found to be completely uniform with no variation whatsoever, then it would have dealt a serious blow to Big Bang cosmology.

If you're having a hard time imagining how stars and galaxies could have "grown" from tiny variations in the thin gas cloud that is thought to have once permeated the cosmos, then there are some helpful analogies. You probably remember growing crystals during elementary school science experiments. If you recall, your starting solution was probably a uniform distribution of dissolved salts, analogous to our thin cloud of primordial gas. In order to grow a crystal structure from this "smooth" solution, you first needed a "seed" crystal, or some kind of surface that initiates the crystal-

---

[19] The free-roaming electrons were captured by the hydrogen and helium nuclei and these gases were the first "official" fully formed elements.

[20] An instant on the cosmic timescale, but it actually took about 100,000 years.

lization process. This "seed crystal" provided a place where the solution was slightly more concentrated, and the laws of nature did the rest. A few days later, a very orderly solid arrangement of chemicals "evolved" from the solutions. It wasn't magic and wasn't a miracle, it was just a necessary consequence of the patterns of providence that we like to call the laws of nature. While the mechanism for growing stars, galaxies, and superclusters is completely different than arranging ions into a crystalline structure, the principle is the same.

In order to test this prediction of cosmic seeds, a special satellite called the Wilkinson Microwave Anisotropy Probe (WMAP) was launched to study the CMBR from outer space.[21] In 2003, the baby picture of the early universe was developed. Just as suspected, the subtle variations in temperature (energy) of the early universe were revealed. From this data, scientists could infer the initial distribution of hydrogen and helium throughout the early universe. Roughly knowing the initial conditions of the cosmic system, they could then apply the known laws of nature, such as gravity, relativity, and nuclear physics to begin reconstructing the amazing chain of events that set the stage for the emergence of our own solar system. This is the science of cosmology.

The Emergence of Celestial Systems

You need more than hydrogen, helium, and trace amounts of lithium to make, planets, plants, animals and people. But after the first few minutes of cosmic history, the universe was no longer hot enough to bake any new material out of these simple ingredients. It would be several hundred million years before nature would start fusing the existing hydrogen and helium into heavier elements such as nitrogen, oxygen, carbon, and calcium. Until that time, gravity would continue to exert its subtle influence on the seeds of cosmic structure, slowly drawing these primordial gas pockets together on different scales, making the first lines of distinction between what would eventually become stars, galaxies, galactic clusters and superclusters.

The highest concentrations of primordial gas first gave birth to supermassive black holes and quasars[22] weighing billions of times more than our

---

[21] The Cosmic Background Explorer (COBE) satellite was launched in 1989 to study the CMBR and was actually the first to detect these subtle irregularities. But its data were not as detailed as the WMAP data.

[22] A quasar is a "quasi-stellar radio source"—a star-like source of light that shines many times brighter than entire galaxies. They are believed to be super massive black-holes inside the nuclei of distant galaxies illuminated by the accretion of material falling into them.

own sun. By the influence of their tremendous gravity, these first structures would lay claim to huge areas of cosmic real estate, carving out what would eventually be the gaseous territory of a future galaxy. Smaller pockets of gas in the surrounding clouds also collapsed in on themselves under the influence of gravity, giving birth to the first stars. These stellar giants slowly made their way into orbits under the gravitational influence of the supermassive black holes. These structures appear today as the spheroid hubs of spiral galaxies, still glowing red from the older stars that remain.

Compared to our own sun, many of these first stars were giants, literally hundreds of times more massive. Stars this size burn extremely fast and extremely hot, lasting only a few million years. All stars "burn" by fusing hydrogen into helium, and these super massive stars were no different. But during their short lives, these giants burned hot enough to fuse together many other elements heavier than helium, all the way up to iron. After quickly spending all of their nuclear fuel, these first stars collapsed in on themselves, resulting in huge explosions that scattered heavier elements out from the galactic hub. These explosions also provided the additional energy needed to "break the iron barrier" and forge heavier elements like silver, gold, lead and uranium.

This process continued for billions of years, enriching the primordial broth with all of the elements of the periodic table, the very things needed for building planets, plants, animals, and people. The ashes of these dying stars can still be seen today as rings of dust living outside the hubs of other spiral galaxies like our own, enriching the younger spiral arms with heavier elements from which other solar systems like ours might arise.

As these clouds of swirling dust and gas took up stable orbits around the galactic hubs like water approaching a drain, they flattened out into rotating disks. As the concentrations of gas and dust within these disks increased, new "seeds" gave rise to smaller stars, forming the galactic "suburbs" outside the hub. Burning slower and not quite as hot as their giant parents and grandparents "downtown," these smaller stars lasted billions of years and most can still be seen burning today. As the newly formed stars were drawn toward the hub by its enormous gravity, the only thing that kept them out in the galactic suburbs was their swirling motion, giving these galaxies their characteristic *spiral* shape.[23]

---

[23] Observations of such galactic structures confirm that the older redder stars are indeed living in the hub, and the younger whiter stars are living on the spiral arms, intermixed with rings of light-absorbing dust, spewed out from the generations of stellar giants that lived and died within the hub. Redshift and blueshift of different stars within the galactic system confirms the swirling motion and allows astrophysicists to calculate the distribution of matter within the galaxy.

Our Solar System

After a while, second and third generation star systems formed from clouds of gas that were enriched with heavier elements from the dying first generation stars. Spectral analysis of light from these younger stars confirms the presence of heavier elements from the recycled interstellar material, again consistent with the theory. About 4.6 billion years ago, on the spiral arm of an ordinary galaxy, an average sized star system began to take shape from one of these enriched gas clouds. As matter continued to accumulate at the center of this cloud, its gravity became stronger and stronger. This in turn pulled in more gas and dust and the pressure at the center of the cloud started to rise. This swirling ball of mostly hydrogen gas with a sprinkling of ashes from its cremated ancestors began to really heat up under the intense pressure. Once the temperature reached 10 million degrees, the thermo-nuclear furnace was switched on and this cloud of dust and gas became our sun!

As the hydrogen fused into helium, the intense heat created enough internal pressure to avoid total gravitational collapse. Its gravity continued to attract more gas and dust and the swirling motion of the surrounding material caused it to flatten out like a rotating disk.[24] The intense solar wind[25] pushed away the lighter elements from the surrounding mix, and the heavier elements that couldn't maintain stable orbits fell into the sun. Eventually, the sun stopped accumulating matter and everything else was in some kind of stable orbit around it. Gravity began to work on whatever was left of the proto-planetary disk, turning dust bunnies into dirt clods, dirt clods into boulders, and eventually boulders into the planets, the moons, and the asteroids we see today.

These objects still bear the cratered scars of their violent past, a time when the smaller chunks of debris were vacuumed up by the gravity of the larger moons and planets until the only things remaining were in some kind of stable orbit around something. Except for the belt of rocks that still remains between Mars and Jupiter and a large quantity of comets, most of the "loose material" has long since been cleaned up by the gravity of the larger moons and planets. The lighter gases pushed away by the solar wind were captured by the enormous gravity of the large outer planets, which became the rocky cores of our solar system's gaseous giants.[26] And there you have the solar system as we know it today.

---

[24] In fact, astronomers have observed this exact phenomenon taking place in the Orion Nebula where enormous clouds of enriched gas are giving birth to stars with similar proto-planetary disks.

[25] Solar wind refers to the more than one million tons of highly charged particles that leave the sun every second at speeds of over 1,000 miles per second.

Completing the Cosmic Puzzle

That wraps up the first nine billion years of cosmic evolution. This brief rundown is woefully simplistic. Many details were omitted, and many details are still unknown. What caused the period of rapid cosmic inflation that smoothed out the distribution of energy within one part in a hundred thousand? What caused the break in symmetry between matter and antimatter of one part in a billion? What caused the sun's proto-planetary disk of dust bunnies to form the half-mile sized boulders necessary for gravity to make large planets? Why do a couple of our planets spin backwards? As with any area of natural science, the more that scientists learn about the cosmos, the more questions remain unanswered. Every time they fill in the gaps with new data, they realize that there are still more gaps. Neil deGrasse Tyson, one of the greatest science writers of our time, sums it up nicely:

> Does any of this sound like the end of science? Does any of this sound like it's time to congratulate ourselves? To me it sounds like we are all helpless idiots...[27]

This tension between what science can explain and what science can't explain creates an interesting situation, like a puzzle with several pieces missing, but with enough pieces in place to see the overall picture.

Are these yet unexplained pieces the result of miracles? They could be, but we need to respond tactfully. When scientists like Dr. Tyson are willing to make honest assessments of their scientific progress, as reflected in the statement above, many Christians smell blood. Rather than enjoy a moment of peaceful reflection with the "opposition" and marvel together at the wonders of God's creation, we reward these genuine acts of humility with what amounts to an I-told-you-so. We see these vulnerabilities as a personal invitation from the scientific community to start waxing theological. I can certainly appreciate the enthusiasm, but if the sovereignty of God is not diminished by the continuous operation of the laws of nature, I see no advantage to filling these gaps with miracles.

For the cosmologists and astrophysicists who spend their lives putting the pieces of this puzzle together, these unknowns represent both frustrations

---

[26] Jupiter, Saturn, Neptune and Uranus are all large gaseous planets with rocky cores. Pluto is too small to gravitationally attract these passing gasses and the solar wind would have pushed them right on by.

[27] Neil deGrasse Tyson, *Death by Black Hole and Other Cosmic Quandaries* (New York, NY; Norton, 2007), pg. 20.

and challenges. However, the unsolved mysteries of creation are not frustrating enough nor are they challenging enough to cause any respectable scientist to completely abandon Big Bang cosmology in favor of supernatural explanations. In science, an answer that can explain anything without risk of being wrong—like a miracle—essentially explains nothing. For them, there are enough pieces of the puzzle already in place to clearly see the "big picture." In fact, every time a new piece is found, it fits perfectly into the rest of the image.

For those that commit their lives to figuring these kinds of things out, there is good reason to believe that someday more pieces will be discovered and the puzzle will be closer to completion. Given the successful track record of astronomy and physics over just this last century, I don't doubt their ability to keep moving forward. This is one of the reasons why supernatural explanations of cosmic history fall on deaf ears within the scientific community. Christians need to be sensitive to this reality and avoid unnecessarily isolating these people from our faith.

Stretching the Imagination

It's perfectly natural to be completely incredulous when hearing the details of cosmic evolution for the first time. How is it possible for natural processes acting on ordinary matter to produce all of this amazing complexity over time? But consider this: if you didn't experience it everyday, wouldn't you also find it hard to believe that natural processes could take a small cell the size of pinhead and grow it into all of the different cells, tissues, and organs of a fully formed human being? Does anybody honestly think biology and chemistry can tell us in exhaustive detail exactly how every step happens along the way? Or how each cell "knows" exactly what to do at exactly the precise moments? Perhaps someday we'll have more of the details worked out, but the natural process of gestation is every bit as incredible as cosmic evolution. Yet once the chain of events is started, it follows its prescribed path to completion, without the need for any "divine intervention" to augment the functional integrity of God's created order.

The human body displays a material complexity that far exceeds anything found in our universe. The inner workings of biochemical processes are shrouded with wonder and mystery. We've only just scratched the surface on all that there is to know and we may never figure it all out. Certainly biologists, doctors, and chemists have a general idea of how gestation progresses from point A to point B, but the entire sequence of events seems to hinge on a tapestry of highly improbable coincidences. It's truly mind boggling. In fact, the birth of a healthy child is so amazing that we often refer to it as the *mira-*

cle of life. Yet, from the random selection of which sperm cell out of the millions trying would make its way to the egg, to the chemical signals that trigger the contractions of the mother's uterus, everything happens according to the wonderful patterns of providence that we call the laws of nature, hundreds of times each day.

The gestation process testifies to the goodness of God and the wonders of creation. Ironically, the same Bible that tells us we are "fearfully and wonderfully made" also tells us that "God created the heavens and the earth." But will Christians ever be able look at the natural development of the cosmos the same way they look at the natural development of a child in its mother's womb? Why is it that some Christians consider the science behind the natural birth of a child a wonderful "miracle," but think that the science behind the natural birth of the universe is just a "godless" theory? Don't the same fundamental forces of nature direct both of these events from start to finish?

Unfortunately, the amazing story of how the cosmos may have developed from its tiny hot beginnings to its present state of enormous complexity can never be directly observed or replayed. It's easy to doubt something that can't be seen. The light received from distant objects is the only information we have that testifies to the amazing natural history of our universe. Because of this, the whole idea of cosmic evolution can stretch our imaginations, especially for scientific laymen like me who don't study these things for a living. But I guarantee you that if the amazing development of a fetus from a zygote to a newborn were not so commonplace, then we would be just as incredulous of it as well.

Cosmic Evolution: Fact or Theory?

Any suggestion that the Big Bang or cosmic evolution could possibly be considered a *fact* is offensive to a lot of folks. The debate is usually focused on what should or shouldn't be taught in public schools. How can anybody be so certain about an event that nobody was around to see and can't even be directly observed today? Good question, but before we start tossing around words like *fact* and *theory*, it might be a good idea to know exactly what these things mean.

The commonly accepted definition of a *scientific fact* is an objective and verifiable observation. A fact is something can be demonstrated to be true or false. A *scientific theory* is a specific explanation or interpretation of the observable facts. A theory that successfully explains the observable facts will stick around until it is shown to be false by a subsequent discovery of contrary facts.

So what about the claim of cosmic evolution? Many of the individual processes such as the formation of stars, galaxies, and solar systems can take millions or billions of years to unfold. It's impossible to directly observe these things happening in our lifetimes. However, the observation of galactic red-shift and the existence of the CMBR are established facts that can be observed, measured, quantified, and verified. As we discussed in Chapter Six, the only coherent explanation of these two facts is that the universe is expanding (redshift) and cooling (CMBR), and was therefore once much smaller and much hotter. If the universe was static and eternal, or shrinking and warming, neither of these things would exist—but they clearly do. Since the definition of cosmic evolution is simply *a change in the structure and properties of the universe over time*, we are absolutely correct to call it a scientific fact. Using our most powerful telescopes to observe deep into the past, scientists can directly observe the evidence of physical change in the structure and properties of the cosmos over time. So based on the galactic redshift, the CMBR and the primitive fringes of the visible universe, cosmic evolution is indeed a scientific fact!

I know this may not sit well with some readers. There is a common misconception that unless you can fully establish a scientific claim beyond a reasonable doubt, you don't have any business calling it a scientific fact. But this is not the case at all. You don't always have to know every detail of how a specific event unfolded in order to know with some certainty that it happened. Again, if you wake up to snow on the ground you can still conclude that it snowed the night before, even if nobody was awake to see it come down. You may never know when it started, when it stopped, how long it fell, or at what rate, but there is enough circumstantial evidence to at least establish the fact that it snowed. And you know enough about how the world works to not believe your neighbor's story about the fleet of levitating snow machines!

It might make some readers feel better to know that cosmic evolution is *also* just a theory. It extrapolates those things that can be directly observed today back to the beginning of time and space in an attempt to provide a coherent explanation of cosmic history. There are many details of exactly how the universe is expanding and cooling that are still very speculative, especially those events that occurred in the distant past. Anything before the first 10^{-11} seconds is almost completely theoretical. I listed a few other unknown details that still boggle the minds of astrophysicists and cosmologists as well. So the bottom line is that like our gravity example from Chapter One, cosmic evolution is *both* a fact and a theory.

It would be similar to the situation of coming home from vacation to find your front door open and some of your belongings missing. The obvious conclusion is that somebody broke into your house and took some of your things.

But what if nobody saw anything and there are no signs of forced entry on your front door? Do you then question the "fact" that you actually got robbed? Of course not. There are other possible explanations of *how* you got robbed, other "theories" if you will. Perhaps the burglar found another way in and left through the front door? Perhaps you forgot to lock the front door and the burglar walked right in? You may never know exactly *how* it happened. The *theory* of how he got inside your home may always remain a mystery, but the *fact* that you got robbed doesn't change. If somebody told you that your stuff miraculously disappeared, you wouldn't consider that a helpful solution to your problem.

And so it is with our explanation of the astrophysical data. There is overwhelming evidence that the Big Bang happened, even though some of the details are still very speculative. Rather than spend all of our time and energy arguing over the details, Christians are better off addressing the spiritual challenges of what this event means to our faith. That's where the scientific rubber really meets the theological road. The cosmologists can have the details. The longstanding belief that the universe is static and eternal has passed. We now have unmistakable evidence that the universe is changing; it is evolving. A necessary consequence of this fact is that the universe also had a beginning, and that beginning was several billion years ago. Many of the details of how all of this fits together are still a mystery. But there is enough conclusive evidence to accept cosmic evolution as a scientific fact that addresses the changing structure and properties of the universe.

CHAPTER EIGHT

THE CHALLENGE OF GEOLOGIC HISTORY

The debate over the age of earth is fueled by the same motivations as the debate over the age of the universe. If the earth is shown to be only a few thousand years old, then evolution would be physically impossible and the debate over the origin of life would be forever won by the Young-Earth Creationists. This seems like a good strategy, but how far are Christians willing to go to try to prove that the earth is only 6,000 years young?

This chapter takes a brief look at the data used by science to determine the earth's age. The most widely accepted number is about 4.55 billion years. Written out, that looks like 4,550,000,000. That's obviously another big number. The biblical estimate for the age of the earth is the same as it is for the age of the universe, around 6,000 to 10,000 years. Since we already looked at the biblical data in Chapter Six, I won't go through it again. But we obviously have another problem. These two estimates are off by a factor of over 75 million percent!

Creation Science and the Age of the Earth

If you've kept up with any creation science material over the years, you've probably seen several "scientific" arguments for the age of the earth. So in addition to the biblical data, which place the earth just a tad over 6,000 years, there is apparently a mountain of scientific evidence that also points to a young earth. In fact, there are at least 68 different ways to "scientifically" conclude that the earth is relatively young, much younger than the 4.55 billion year estimation of science. So why do scientists continue to suppress this "overwhelming" evidence for a young earth? Are they so motivated by their opposition to biblical Christianity that they won't consider any legitimate scientific arguments for a young earth, or is there something else going on here? Let's take a brief look at the nature of these Young-Earth arguments and try to honestly understand why they have been universally rejected by the mainstream scientific community.

The following table lists each of these 68 Young-Earth arguments and their corresponding age estimates. In fairness, I should note that the list is over 20 years old, but since these same evidences continue to circulate through creationism literature to this day, the list is still relevant to our present discussion. There may be a few other arguments that are not included on this list. In fact, I know of a couple that have been "discovered" since this list was originally put together. So if you can't find your personal favorite here, it's probably because it is fairly new, so I apologize in advance. But I'm pretty confident that if this list went from 68 to 69 or 70, the overall interpretation of the data wouldn't change much.

You can be certain of this: any conclusive evidence for a young earth would be about as revolutionary as any conclusive evidence for a change in the speed of light. If any one of these 68 arguments were convincing enough to meet the minimum standards of scientific scholarship, the news of it would be all over the place. I should also note that while this list has been used by many on both sides of the creation/evolution debate, it was originally tabulated by two leaders in the Young-Earth Creationism movement as undeniable proof that the earth can't be as old as science claims that it is.

Table 1: Sixty-Eight Common Young-Earth Proofs[1]

	Young-Earth Argument	Age of Earth
1.	Decay of the earth's magnetic field	10,000
2.	Influx of radiocarbon to the earth system	10,000
3.	Influx of meteoritic dust from space	Too small
4.	Influx of juvenile water to the oceans	340,000,000
5.	Influx of magma from the mantle to form the crust	500,000,000
6.	Growth of oldest living part of the biosphere	5,000
7.	Origin of human civilizations	5,000
8.	Efflux of Helium into the atmosphere	1,750–175,000
9.	Development of the total human population	4,000
10.	Influx of sediment to the ocean via rivers	30,000,000
11.	Erosion of sediment from the continents	14,000,000
12.	Leaching of sodium from the continents	32,000,000
13.	Leaching of chlorine from the continents	1,000,000
14.	Leaching of calcium from the continents	12,000,000
15.	Influx of carbonate to the ocean	100,000

---

[1] H. M. Morris & G. E. Parker, *What is Creation Science?* (San Diego, CA; Creation-Life, 1982), Table I, pg. 254-255.

16.	Influx of sulfate to the ocean	10,000,000
17.	Influx of chlorine to the ocean	164,000,000
18.	Influx of calcium to the ocean	1,000,000
19.	Influx of uranium to the ocean	1,260,000
20.	Efflux of oil from traps by fluid pressure	10,000–100,000
21.	Formation of radiogenic lead by neutron capture	Too small
22.	Formation of radiogenic Sr by neutron capture	Too small
23.	Decay of natural remnant paleomagnetism	100,000
24.	Decay of carbon in Precambrian wood	4,000
25.	Decay of uranium with initial "radiogenic" lead	Too small
26.	Decay of potassium with entrapped argon	Too small
27.	Formation of river deltas	5,000
28.	Submarine oil seepage into the ocean	50,000,000
29.	Decay of natural plutonium	80,000,000
30.	Decay of lines of galaxies	10,000,000
31.	Expanding interstellar gas	60,000,000
32.	Decay of short-period comets	10,000
33.	Decay of long-period comets	1,000,000
34.	Influx of small particles to the sun	83,000
35.	Maximum life of meteor showers	5,000,000
36.	Accumulation of dust on the moon	200,000
37.	Instability of the rings of Saturn	1,000,000
38.	Escape of methane from Titan	20,000,000
39.	Deceleration of the earth by tidal friction	500,000,000
40.	Cooling of the earth by heat efflux	24,000,000
41.	Accumulation of calcareous ooze on the sea floor	5,000,000
42.	Influx of sodium to the ocean via rivers	260,000,000
43.	Influx of nickel to the ocean via rivers	9,000
44.	Influx of magnesium to the ocean via rivers	45,000,000
45.	Influx of silicon to the ocean via rivers	8,000
46.	Influx of potassium to the ocean via rivers	11,000,000
47.	Influx of copper to the ocean via rivers	50,000
48.	Influx of gold to the ocean via rivers	560,000
49.	Influx of silver to the ocean via rivers	2,100,000

50.	Influx of mercury to the ocean via rivers	42,000
51.	Influx of lead to the ocean via rivers	2,000
52.	Influx of tin to the ocean via rivers	100,000
53.	Influx of aluminum to the ocean via rivers	100
54.	Influx of lithium to the ocean via rivers	20,000,000
55.	Influx of titanium to the ocean via rivers	160
56.	Influx of chromium to the ocean via rivers	350
57.	Influx of manganese to the ocean via rivers	1,400
58.	Influx of iron to the ocean via rivers	140
59.	Influx of cobalt to the ocean via rivers	18,000
60.	Influx of zinc to the ocean via rivers	180,000
61.	Influx of rubidium to the ocean via rivers	270,000
62.	Influx of strontium to the ocean via rivers	19,000,000
63.	Influx of bismuth to the ocean via rivers	45,000
64.	Influx of thorium to the ocean via rivers	350
65.	Influx of antimony to the ocean via rivers	350,000
66.	Influx of tungsten to the ocean via rivers	1,000
67.	Influx of barium to the ocean via rivers	84,000
68.	Influx of molybdenum to the ocean via rivers	500,000

These are 68 different natural processes that creation science uses to cal-culate the earth's age. And now I will offer a point-by-point evaluation of the specific details for all 68 arguments. Not really; I wouldn't subject any reader to that. The information is out there if you're really interested. Just *Google* any one of these arguments and you'll find more than enough discussion for and against each one. But does it really even matter? Look at these numbers again. Does anything seem strange to you? The biblical account supposedly tells us that the earth is just over 6,000 years old, but not one of these other "proofs" are even within 1,000 years of that! The rest of them give ages that range from 100 years to 500 million years. The average of these ages is 3.4 million years; the median age is 200,000 years. Statistically speaking, these data are completely useless for determining anything!

I hate to be critical of other Christian brothers, but this is flat-out embar-rassing. There is nothing wrong with trying to find evidence that supports the biblical claim of a 6,000 year-old earth in the form of scientific data, but these 68 different processes can't tell you anything conclusive. If a scientist were honestly searching for different ways to objectively date the age of the earth without any preconceived ideas, they would quickly realize that these 68 processes are not reliable enough to draw any meaningful conclusions. In fact, the extreme variance in the data can be easily explained in terms of other natural processes that have nothing to do with the age of the earth.

The more revolutionary that a scientific claim is, the more solid the data should be that support it. It's not impossible to turn the entire science of geology upside down, but it's going to take better data than these. Even a genius like Einstein didn't try to overturn the popular conception of a static universe with his cosmological equations alone. It required Hubble's evidence of cosmological redshift to begin the long process of scientific revolution, and it took 40 years to complete.

That's just the way science works. The system is not rigged against Christians any more than it is rigged against any other person with a novel idea. The same internal "checks and balances" apply to anybody who wishes to challenge the existing paradigms. We Christians can't just expect to turn the world upside down with incoherent data like these. Who do we think we are? Does being Christians and having the Bible exempt us from the basic standards of scientific conduct? Do we think that scientists will take anything we say seriously when we commit these flagrant violations of academic scholarship and scientific protocol?

So why do creation scientists keep using these 68 young-earth arguments if they have all been dismissed by the mainstream scientific community? The answer can be found in their own interpretation of the data:

> There are, as a matter of fact, scores of worldwide processes which give ages far too young to suit the standard Evolution Model. There are 68 types of such calculations listed in Table I, all of them independent of each other and all applying essentially to the entire earth, or one of its major components or to the solar system. All give ages far too young to accommodate the Evolution Model.[2]

These 68 natural processes were specifically chosen by Young-Earth scientists for the sole purpose of arguing *against* a 4.55 billion year-old earth, not to establish a scientific basis for accurately determining its age.

This is no different than a criminal defense attorney who has no way to explain the obvious facts of the case against his client. His only remaining strategy is to muddy the waters with a bunch of red herrings, to create confusion and reasonable doubt in the minds of the jury by gumming up the works

---

[2] Ibid, pg. 251.

of the prosecution's case. But these questionable tactics apparently don't bother the creation scientists. They don't need coherent data to determine the age of the earth; they just need something to demonstrate that "science" can prove the other side wrong. The Bible tells them how old the earth is, these 68 other data points are just used to show that earth can't be as old as most scientists say that it is. This is not how science is supposed to work! The data have to support your claim. It's not enough just to contradict the other guy's idea with a mess of statistically incoherent numbers.

The Problem with Young-Earth Arguments

If you search the world over, you can find a multitude of natural processes that ebb and flow through the natural course of history (apparently there are at least 68 of them). These "worldwide processes" are mostly very complex phenomena that are dependent on many other natural processes. Some are very well understood and others are not. It is somewhat naïve and opportunistic to uniformly extrapolate a relatively small sample of data backwards in time from any of these current trends and draw a conclusion without more information.

For example, if I went to the beach at low tide, observed a four foot rise in the sea level over a six-hour period, and then published a "scientific" paper concluding that in just over five years the tallest mountains will be swallowed by the sea based on my detailed observations, nobody would take me seriously. Why? Because everybody knows that the rising and falling tides are complex processes with natural cycles and frequencies. Each of these 68 natural processes is also extremely complex and, like the rhythm of the ocean's tides, involves many other natural phenomena that cause them to "wax and wane" as the tides do.[3]

Trying to accurately determine the earth's age from any one of these 68 natural processes is like trying to measure global warming by watching a thermometer for a few hours. If these 68 natural processes were valid methods that could be used to accurately determine the age of the earth, the data would tend to converge around a single value. There would be a coherent distribution with an obvious mean value and a reasonable deviation that could be explained by other factors. In the best case scenario, the convergence would be consistent with the biblical data of a 6,000 year-old earth, but a convergence around any value would at least indicate some statistical significance in the data. Unfortunately, these 68 arguments do no such thing and therefore the

---

[3] Not all 68 processes are necessarily cyclical like the tides. But the point is that none of them have been shown to be uniform over extended periods of time.

mainstream scientific community is perfectly justified in not using any of them to determine the age of the earth.

Regardless of the obvious problems with these 68 young-earth arguments, the leaders of the creation science movement summarize them by the following statement:

> Thus, it is concluded that the weight of all the scientific evidence favors the view that the earth is quite young, far too young for life and man to have arisen by an evolutionary process. The origin of all things by direct creation—already necessitated by many other scientific considerations—is therefore also indicated by chronometric data.[4]

Are we to infer that these 68 data points constitute the "weight of all scientific evidence" pertaining to the earth's age? I understand the motivation to try to uphold the inerrancy of the Bible, but this desire is misguided and these tactics are counterproductive. Such lack of scientific integrity should be the means to no spiritual end. These are nothing but 68 reasons why you can't believe any scientific claims of the Young-Earth Creationists. Unfortunately they also backfire on Christians in general, quickly becoming 68 reasons why mainstream science will never take the Bible seriously. If we are going to treat science as a mission field, we must do better than this.

Science and the Age of the Earth

So what processes are used by the secular scientists to date the age of the earth? Obviously, there must be something that avoids the same pitfalls as the 68 Young-Earth arguments. As we have seen, the only way to get consistent data is to use natural processes that have a definite starting point, a steady rate of change that is independent of environmental effects, and a definite ending point, like starting and stopping a stopwatch. If only somebody had just started their stopwatch when the planet formed and let it run, we could just take a peek at it and the age of the earth would be revealed. Since nobody was around to do that, we'll just have to be a little more innovative.

Fortunately for us, nature provides us with a process that is just as steady as any time piece. It is called *radioactive decay* and it works like this: certain isotopes[5] found in nature are unstable. These individual atoms, for whatever reason, have a nucleus that doesn't want to keep its identity. It will eventually

---

[4] Ibid, pg. 252.

get rid of some of its contents and become something else. This process is called radioactive decay because the original isotope (the parent) "decays" into something else (the daughter) by "radiating" away some of its contents.

The great thing about radioactive decay is that any sample of an unstable isotope will decay at a very precise rate. The rate is determined by the strong and weak nuclear forces and paces itself by the speed of light. Unlike many of the other processes favored by the Young-Earth scientists, radioactive decay is totally independent of the environment. Hot or cold, light or dark, wet or dry, on earth or in space, it doesn't matter. The rate of nuclear decay for any unstable isotope is as steady and as unchanging as the speed of light through empty space.

The rate at which a sample of any unstable element decays is measured by its *half-life*. For example, a particular isotope of uranium (^{235}U) has a half-life of 704 million years. That's just another way of saying that after 704 million years, half of this stuff will decay into a particular isotope of lead (^{207}Pb), and half will remain uranium.[6] After 704 million more years, the half of uranium that remains will decay to only a fourth of the original material. This process keeps on going until there is practically nothing left. But the original sample of ^{235}U will never completely disappear, it will just keep getting cut in half every 704 million years until whatever is left becomes too small to detect. Once that happens you can pretty much say that it's gone.

So if you start out with a known quantity of the parent isotope, you can determine the age of the sample with incredible precision just by measuring how much of it has decayed into the daughter isotope and applying some basic math. How hard is that to do? Nowadays, it's not hard at all. Analyzing the amount of any specific element in a given sample is easily done, calculating the age of the sample is fairly simple; the only challenging thing is to know how much of each material you started with. For example, if a rock was formed billions of years ago with a certain amount of parent and daughter isotopes, and nothing was added to or removed form the rock since, you could tell exactly how old it is by measuring the amount of parent and daughter isotopes in the rock today. Again, the challenge is to know what the original quantities were, and to make sure that your samples didn't get compromised after their formation. Fortunately for us, there are some pretty straightforward ways to do this that we'll look at later. But first, we can apply an even easier test of the earth's age by radioactive decay.

---

[5] Almost every element has isotopes. They are chemically similar versions of each other that differ only by the number of neutrons on their nucleus.

[6] Actually, ^{235}U goes through a few intermediate steps before eventually getting to ^{207}Pb. The complete process has a half-life of 713 million years.

Twenty-Nine Reasons Why the Earth is Old

In addition to ^{235}U, there are hundreds of other radioactive isotopes that all have their own half-lives. Most of the half-lives are relatively short so they don't do us much good if we're talking about tracking them over millions of years. A few of them that have longer half-lives are continually being reproduced by other processes which make it hard to determine how much of the original material has decayed. So if we only consider radioactive isotopes with half-lives longer than a million years, there are about 29 of them that are not re-created by other natural processes.[7] In other words, once they decay they are gone forever.

If you recall from the last chapter, all elements heavier than iron, including all but two of our 29 radioactive isotopes, were forged from the tremendous energy and heat of exploding stars called supernovae. As soon as these unstable elements were created, they were dumped back into interstellar space and the process of radioactive decay took over. It is from one of these enriched clouds of primordial hydrogen and helium that our solar system is believed to have formed. So scientists should be able to get a rough idea of how old our solar system is by detecting what isotopes are still hanging around in measurable quantities. For instance, if all 29 are still here, then our universe can't be older than a few tens of millions of years. If none of them can be found, then our universe must be practically eternal.

As it turns out, of these 29 radioactive iso opes that would have been present during the formation of our solar system, 11 of them can not be found anywhere in nature, at least not in our corner of the galaxy. Why is that? Why is it that these 11 isotopes are completely missing from our solar system but the other 18 isotopes are all present and accounted for? When y u look at the half-lives of each of the 29 isotopes, the trend becomes clear. Everything with a half-life of less than 80 million years has disappeared completely from our solar system. Everything with a half-life greater than 80 million years still remains. What does that mean?

A half-life of 80 million years means that for a given sample of radioactive material, half of it will still remain 80 million years later. After 160 million years, a quarter of it will remain. After 240 million, an eighth will remain, and so on and so forth. This process continues until there is practically nothing left of the parent isotope that can be detected. For something that started

---

[7] I first ran across this argument while reading *Finding Darwin's God: A Scientist's Search for Common Ground Between God and Evolution* by Kenneth R. Miller (New York, NY; HarperCollins, 1999), pg. 69-72. The original data was taken from *The Age of the Earth* by G. Brent Dalrymple (Stanford, CA; Stanford University Press, 1991), pg. 377.

out in abundance to become completely undetectable in nature, nuclear physicists say that it has to decay through at least 20 half-lives. That means for something with a half-life of 80 million years, it would take at least 1.6 billion years to "disappear."

So what are we to conclude from this? Obviously this can't give us the absolute age of our solar system or our earth because not enough information is known about exactly how much of what isotopes we started with and where they came from, but it can give us some general information. For example, our solar system has not been around forever. That much is clear. This may seem like trivial information, but none of these isotopes would be present if our solar system had existed from eternity.

Another important observation is that our solar system must be at least a *billion* years old. If it were any younger than that, we should be able to detect some of the radioactive isotopes with half-lives shorter than 80 million years. For example, if our solar system were only 6,000 years old, we would be able to easily find all 29 of our radioactive isotopes because they all have half-lives of over 1 million years. So even this simple example of radioactive decay is enough to estimate the earth's age at over a billion years. But can we get any more specific than this?

The Stones Will Cry Out

In order to obtain reliable results for the age of the earth using this method, four things are needed. First, you need a sample that contains a radioactive isotope with a known half-life, like an old rock taken from the earth's crust. Second, you need to be able to measure the relative amounts of parent and daughter isotopes contained in your sample. Third, you need to make sure that your sample has not been compromised since some geological processes can alter the quantities of parent and daughter isotopes in your sample and give you false readings. Fourth, you need to know the relative amounts of parent and daughter isotopes that were present when the rock was formed. Once you have all four of these things, you simply take your measurements, apply a little simple math, and out comes the age of your rock! Does this sound challenging? Some might have you believe that knowing these four things is next to impossible. Let's look at them one at a time.

A rock is a rock. This may sound like the easiest one of the four steps. Just go into your backyard and pick one out of the ground, right? Not so fast. Provided that you could accurately obtain the age of *that* particular rock, it still probably won't tell you the overall age of the earth. In order to know when the earth was formed, you would need a rock that was there from the very beginning, which might be a problem. Any melting down and reforming of the earth's crust would have basically "reset" the radiometric clocks. If the early earth was as geologically active as many scientists think it was, this activity could have "erased" the data from the earliest samples. If that is true, then the oldest rocks obtained from the earth would really only tell us how long ago the earth became geologically stable, not when the earth was formed. Unfortunately, these same geological processes would have affected every other planet and the moon as well, so sending space probes out into the solar system to retrieve extraterrestrial rocks isn't going to help either. The only rocks in our solar system that would have been unaffected by geological activity are the asteroids.

The second requirement was that we be able to measure the relative amounts of parent and daughter isotopes in the sample. This is a fairly straightforward process that is made possible by modern technology. Even trace amounts of isotopes can be accurately detected.

The third and forth concerns were that the original sample is not compromised by geological processes, and that the original amounts of both parent and daughter isotopes are known. I lumped these two things together because they are both solved by using the same dating method, otherwise known as *isochron* dating.[8] Here is how it works: a typical rock is usually made up of several different kinds of minerals. Each mineral contains various types and quantities of radioactive isotopes. When the rock is first formed, it starts out with a certain amount of parent isotope (P) and daughter isotope ($D+d$) "locked" into the various different minerals that make up that rock. Some of the original daughter isotope would have been *primordial daughter* (D), and some of it would have been *radiogenic daughter* (d).[9]

Unfortunately, there is no way to directly determine what the original quantities of P, D or d would have been for each of the rock's minerals. But what's so amazing about the isochron method is that you don't even need this information. You only need to know how much P, D and d you have now. How is this possible? Does this not sound like a perfect opportunity for biased scientists to manipulate the data? Actually, quite the opposite is true. The isochron method takes the human bias right out of the process.

---

[8] My explanation of isochron dating is over-simplified for non-technical readers. If you want the highly technical version in exhaustive detail, I recommend *The Age of the Earth*, by G. Brent Dalrymple (Stanford, CA; Stanford University Press, 1991). Or you can *Google* "isochron dating" and find more than enough information.

[9] *Primordial* daughter isotopes were originally produced by stars. *Radiogenic* daughter isotopes were produced from the radioactive decay of the Parent isotope since its creation.

Even though the original amounts of P, D and d can't be determined directly, the relationships between them can be logically deduced from what is presently known. For instance, since different types of minerals incorporate different chemical elements in various quantities, the ratio of original parent isotope to primordial daughter (P/D) would have been *different* for each of the rock's minerals. That's obvious enough. On the other hand, the original ratio of radiogenic daughter isotope to primordial daughter isotope (d/D) would have been the *same* for each mineral. That might not be quite as obvious, but it is a safe assumption for two reasons. Firstly, the entire rock, including all of the various minerals that it contains, would have been formed from a homogeneous elemental mix of d/D. And secondly, since D and d are different isotopes of the same element and are therfore chemically identical, each of the rock's minerals would have incorporated them by the same ratio of d/D that was in the overall elemental mix that formed the rock.

So if we plot P/D versus d/D for all of the minerals *initially* present in the rock and connect the dots, it would look like a horizontal line somewhere on the chart. Why? Because for each mineral, P/D would have been different and d/D would have been the same.[10]

Once the rock was formed, the original quantities of P, D and d would have been fixed along that line, somewhere on the chart. Over time, the parent isotope (P) in each individual mineral would start to decay into the radiogenic daughter isotope (d). Now fast-forward a few billion years. The *increase* in d will be identical to the *decrease* in P, since P only decays into d. So the minerals in the rock with more P (towards the right of the chart) will end up with more d (higher on the chart). The minerals in the rock with less P (towards the left of the chart) will end up with less d (lower on the chart). And the original amount of D doesn't change because it neither decays, nor is made by decay. So the measured quantities of P, D, and d will all lie along a straight line with a positive slope. And the slope of the line allows you to directly calculate the age of the sample.[11] Figure 4 illustrates how isochron dating works.

---

[10] This assumption is self-correcting. If for some reason d/D is not identical for each mineral, the measured data will not be co-linear and the sample is discarded.

[11] If enough time were to pass to allow all of the P to decay into d, the line would be vertical.

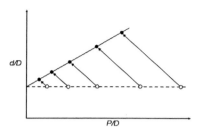

Figure 4: This is an example of isochron dating for a hypothetical rock with five different minerals. Each mineral started out with the same unknown ratio of d/D, but with a different unknown ratio of P/D. The original unknown data are represented by the hollow dots and the dotted line. The solid dots represent the measured ratio of d/D and P/D. The solid line drawn through the laboratory data intersects the vertical axis at the original d/D value. Knowing this and the slope of the solid line allows the age of the sample to be calculated without ever having known exactly what the original amounts of P, D and d were.

The isochron method kills two birds with one stone! It takes care of the unknown original quantities of P, D and d, and it also tells you when your sample has been compromised. If the data from each mineral in your sample are not co-linear, then one of the assumptions is false and you toss out the sample. The isochron method allows scientists to only date rocks whose minerals are capable of telling a coherent story. Each mineral in the rock has to "agree" on what the rock's age is. If any one of them is "out of line" the sample is not used.

Just like that, scientists can take a rock composed of various minerals and estimate the time since the minerals were formed just by measuring the current quantities of parent and daughter isotopes. The process is really an elegant and beautiful thing. You don't need to know the original composition of the sample and the process even tells you if it has been compromised. Whether you like the results or not, you have to admit that the isochron dating method is pretty impressive.

According to the theory of how our solar system evolved from a swirling disk of dust and gas, all of the planets, moons, and asteroids should have been formed within a few hundred million years of one another—practically a simultaneous event on the vast cosmic timeline. However, the catastrophic geological activity on all newly formed planets and moons would have reset the radiometric clocks by dropping the slope of the isochron lines back to zero (horizontal). Therefore, the oldest rocks from these sources should consistently give younger ages than the actual age of our solar system.

As it turns out, the oldest rocks from planet Earth consistently give ages between 3.5 and 3.8 billion years using five different isotope decay processes. Because of the catastrophic geological history that was believed to have marked the first billion years of the earth's history, most scientists assume that these ages are a more accurate reflection of the final cooling of the earth's crust rather than the age of the earth itself. Of the hundreds of moon rocks brought back from various lunar missions, samples from Apollo 16 and 17 give ages between 4.0 and 4.5 billion years. Rocks from the lunar lowlands give ages between 3.5 and 4.0 billion years.

Despite this powerful tool of geological science, isochron dating methods can be subject to misinterpretation. For instance, the isochron line of an uncontaminated sample always starts out horizontal. Radioactive decay of *P* into *d* gives the line a positive slope. If the ratios of *P/D* or *d/D* are compromised, the data will not be linear and the sample can not be used. However, if a catastrophic event destroys and reforms the rock, such as volcanic melting and cooling, it will reset the isochron line to the horizontal position and there will be no direct indication of compromise. So any error in isochron dating methods always gives *younger* ages, not older ages. Therefore, it is believed that the wide variation found in samples taken from the earth and moon is due to these types of catastrophic geological processes, and the actual age of the solar system is *older*.

According to this scenario, the only logical choice for truly dating the age of our solar system is the asteroids. So what about the asteroids that fall to Earth as meteorites? If they are not subject to these kinds of geological processes, how old do they appear to be? Table 2 is a list of over a hundred samples from a dozen meteorites using five separate radioactive isotopes.

Table 2: The ages of various meteorites found on Earth indicate that our solar system is about 4.55 billion years old.[12]

Meteorite	Dated	Method	Age (billions of years)
Allende	whole rock	Ar-Ar	4.52 +/- 0.02
	whole rock	Ar-Ar	4.53 +/- 0.02
	whole rock	Ar-Ar	4.48 +/- 0.02
	whole rock	Ar-Ar	4.55 +/- 0.03
	whole rock	Ar-Ar	4.55 +/- 0.03
	whole rock	Ar-Ar	4.57 +/- 0.03
	whole rock	Ar-Ar	4.50 +/- 0.02

---

[12] G. Brent Dalrymple, *The Age of the Earth*, (Standford, CA; Stanford University Press, 1991), pg. 286.

Guarena	whole rock	Ar-Ar	4.44 +/- 0.06
	13 samples	Rb-Sr	4.46 +/- 0.08
Shaw	whole rock	Ar-Ar	4.43 +/- 0.06
	whole rock	Ar-Ar	4.40 +/- 0.06
	whole rock	Ar-Ar	4.29 +/- 0.06
Olivenza	18 samples	Rb-Sr	4.53 +/- 0.16
	whole rock	Ar-Ar	4.49 +/- 0.06
Saint Severin	4 samples	Sm-Nd	4.55 +/- 0.33
	10 samples	Rb-Sr	4.51 +/- 0.15
	whole rock	Ar-Ar	4.43 +/- 0.04
	whole rock	Ar-Ar	4.38 +/- 0.04
	whole rock	Ar-Ar	4.42 +/- 0.04
Indarch	9 samples	Rb-Sr	4.46 +/- 0.08
	12 samples	Rb-Sr	4.39 +/- 0.04
Juvinas	5 samples	Sm-Nd	4.56 +/- 0.08
	5 samples	Rb-Sr	4.50 +/- 0.07
Moama	3 samples	Sm-Nd	4.46 +/- 0.03
	4 samples	Sm-Nd	4.52 +/- 0.05
Y-75011	9 samples	Rb-Sr	4.50 +/- 0.05
	7 samples	Sm-Nd	4.52 +/- 0.16
	5 samples	Rb-Sr	4.46 +/- 0.06
	4 samples	Sm-Nd	4.52 +/- 0.33
Angra dos Reis	7 samples	Sm-Nd	4.55 +/- 0.04
	3 samples	Sm-Nd	4.56 +/- 0.04
Mundrabrilla	silicates	Ar-Ar	4.50 +/- 0.06
	silicates	Ar-Ar	4.57 +/- 0.06
	olivine	Ar-Ar	4.54 +/- 0.04
	plagioclase	Ar-Ar	4.50 +/- 0.04
Weekeroo Station	4 samples	Rb-Sr	4.39 +/- 0.07
	silicates	Ar-Ar	4.54 +/- 0.03

As you can see, radiometric dating of several isotopes using the isochron method gives a very consistent picture. The data are extremely coherent and are statistically significant. Compare these data with the 68 natural processes used by the leading Young-Earth Creationists to establish the age of the earth. Need I say more?

The Christian Response

As was the case for the data indicating the antiquity of the cosmos, Christians can respond to the data for an old earth in a variety of ways. Again, if we assume that the Bible is giving us creation science, then we either have to find a way to explain away the apparent 4.55 billion year history of the earth, or we must find a way to read the Bible so that it actually supports the idea of an old earth. This is another tall order for the creation scientists. Unfortunately, it too divides the church over nonessential doctrines. Rather than working together to liberate science from its philosophical bondage to materialism, YECs and OECs just end up fighting with each other over how to apply the Bible to the study of natural history. The secular world could easily point to this spirited infighting and declare that Creation Science is a "theory in crisis."

Young-Earth Creationists, assuming that the earth must be six to twelve thousand years old, have no choice but to explain away the geophysical data. One way to do this is by again using the "appearance of age" argument. This same line of reasoning is extended from the light of distant celestial objects to include rocks and meteorites created with the proper amounts of parent and daughter isotopes so the "divine charade" of natural history is accurately maintained. Apparently for those Christians who espouse this viewpoint, there is no end to how far God will go to trick people into thinking that the earth is old.

Another way to argue the case for a young earth is to continue pushing the 68 (or more) discredited Young-Earth arguments. This might work on other Christians who don't have the scientific tools in the toolbox to properly evaluate these arguments, but the world will not be impressed. Put yourself in the shoes of a secular scientist. Pretend for a moment that you have no particular religious or philosophical commitment to the idea of a young earth. Now suppose that you are confronted with two different sets of data for determining the age of our planet. On the one hand, you have 68 different natural processes that all take place at various rates. These processes are very complex, many of the rates are not well understood and there is absolutely no way to determine if the rates observed today are consistent with past rates, or if they are cyclical with extremely long periods that we have yet to realize.

Regardless of the obvious complications, you look at the data anyway only to find that these 68 processes give ages for the earth that range from 100 years to 500 million years. But the person showing you data insists that the true age is 6,000 years, despite the fact that not one of the 68 arguments is even within 15% of that! Are you feeling confident about this data set yet?

On the other hand, you are shown over 100 different samples from dozens of meteorites that were independently dated in different laboratories using a natural process with rates as steady as an atomic clock that are not affected in any way by the environment. The methods used are relatively straightforward and have a built-in error-control mechanism that allows you to toss out contaminated samples (only colinear data are reliable). This approach produces a statistically significant set of data with over 100 possible data points all tightly converging toward a single age of 4.55 billion years with a very tiny deviation. Now who are you supposed to believe?

Is it any wonder why scientists don't take Christians seriously? These kinds of embarrassing things will happen when the scientific method gets hijacked by predetermined conclusions before an honest investigation of the facts is completed. This dogmatic commitment to a young earth forces Christians into giving strange explanations of the scientific data. Since everything in science is tentative, including the commonly accepted age of the earth, I'm not going to rule out some future discovery that conclusively demonstrates that the earth is young. But for now, you're not likely to see any of the current Young-Earth arguments in peer-reviewed scientific publications. There is simply nothing scientific about them.

This is not an anti-Christian conspiracy as the YECs claim; it's just how science works. Anybody is free to offer evidence that challenges the existing paradigms. No matter how good a scientific theory is one day, it can always be challenged the next—providing there is some kind of evidence that supports the new claim. Einstein was a clerk in the Swiss Patent Office when he turned the world of physics upside down. In fact, most of the Nobel Prizes that get handed out go to scientists who were once considered the underdogs, fighting to overthrow the scientific establishment with a new discovery.

That being said, there are certain rules and procedures that must be followed in order to go about something like this. Of all people, Christians should be the first to work within the ethical standards of any profession, but when it comes to Young-Earth Creation science, this is sadly not the case. The YECs, convinced that the mainstream scientific community has conspired against them, purposely and flagrantly violate the basic ethics and protocols of scientific scholarship. As a result, they often find themselves in the crosshairs of other scientists who take the standards of their profession seriously. In fact, some of the best critics of YEC scientists are the OEC scientists who are tired of getting beat-up by their secular colleagues over the ridiculous claims and underhanded tactics of their Young-Earth brothers.

The other tactic is to attack the methods and accuracy of radioactive decay. Of course no process is 100 percent foolproof 100 percent of the time, and radiometric dating is no exception. There will always be errors in some measurements and some samples will undoubtedly give you false readings. If you spend any time looking into this, you will find YEC scientists trying to capitalize on these rare anomalies and create reasonable doubt in the mind of the public. But the evidence for a 4.55 billion year old solar system is not merely based a few selected samples. Look at Table 2 again. When you literally have hundreds of data points converging on a single value with a very small error, it's not too hard to explain a few anomalous readings. It would be unreasonable to discard the entire process because of a few questionable measurements.

One of the more ironic charges against radiometric dating is the accusation that geologists only record data for samples that "agree with evolutionary conclusions." Anybody with even the most basic understanding of how radiometric dating works can tell you that some samples are indeed tossed out and some data are discarded. So is this just the result of evolutionary bias in the laboratories? Is secular geology busted? Have they been caught fudging the facts? Has the Old-Earth conspiracy been exposed? Not quite.

If every single mineral in a sample doesn't give you the same age, then the obvious conclusion is that the sample was compromised and the whole thing gets tossed out. Since corrupted samples sometimes give younger ages, some data indicating a younger earth have to be discarded. However, this is not a result of evolutionary bias; this is actually the result of a very high standard for sample integrity in the radiometric dating process that discriminates against incoherent data. If only the YEC scientists would have the same checks and balances in their "laboratories" we wouldn't have these other 68 embarrassing arguments that are supposed to prove beyond a shadow of a doubt that the earth is only 6,000 years old.

When confronted with this evidence, many young-earth Christians wrongly assume that this is all a big conspiracy against creationism. Let me first say that this is a "great" way to build bridges with the scientific community. Hundreds of creationist books and videos have been blaming the conclusions of mainstream science on dishonesty and deception. I understand the motivation, but these claims are simply absurd. Just think about it for minute. The worst thing for a scientist's career is unoriginality. Nobel Prizes are not handed out to those who simply agree with the establishment. In order to make a name for yourself, you need to buck the system. You need to uncover something that nobody else has considered and turn the establishment upside down.

Determining the age of the earth is no different. If there were any shred of evidence that could possibly show that the earth is young, religious considerations would quickly fall by the wayside. The potential fame and fortune for anybody who can prove a young earth is simply too great to pass up. Who wouldn't want to start a revolution in geology!

In a way, science is like capitalism. Self-interest and competition is what makes the entire system work. Everybody wants to be "that guy" who makes the next big discovery. Conversely, nobody is going to let somebody else take credit for a new discovery unless it totally checks out first. That is one reason why there is such an extensive peer-review system.[13] As a result, scientists are always looking over each others' shoulders trying to find out what the other guys are up to. If something new is proposed, the other groups that are working on the same problem immediately test, retest, check and recheck the new idea to find anything that could have been overlooked.

In this kind of atmosphere, it would be impossible to maintain a worldwide conspiracy that is able to successfully suppress legitimate evidence for a young earth. The only reason that all of these laboratories around the world agree on the age of the earth is because all of the evidence, as we know it today, is conclusive. There may not be anything too "exiting" about all of these scientists independently coming to the same conclusion, but sometimes that happens.

Most Christians who spread these "conspiracy" rumors are not even part of the scientific community. Anybody who has worked closely with the scientific establishment knows how religious implications are rarely considered. For them, it's all about the facts and data. Consider the personal testimony of a Christian who studied physics in order to be a "champion of scientific creationism," but instead became a strong advocate against Creation Science:

> …something happened en route to my Ph.D. I began to see that modern science was not the grand hoax the creationists said it was. In fact, I discovered that science was ruthlessly honest and that this honesty was at the very heart of the scientific process. True, there have been hoaxes perpetrated by individuals within the scientific community, but it was this community, policing itself in the interest of truth, that detected and exposed the hoaxes. I discovered that philosophical and religious considerations were not a part of the scientific method and that most scientists were quite incompetent in matters of philosophy and religion, which they perceived as irrelevant to their disci-

---

[13] The other reason for the peer-review system is to allow other scientists to evaluate scientific claims before making them public. Since most people do not have the ability to evaluate the accuracy of scientific claims, it is considered highly unethical to bypass the peer-review process and go straight to the public with a new idea. Yet, walk into any Christian bookstore and you'll find just that.

pline.

Scientists working in laboratories, at computers, and in the field were, it seemed to me, gathering little bits of truth called facts, which they would then explain in the most logical way possible. Scientists seemed to pay absolutely no attention to the religious implications of their work. Try as I might, I could not find this coven of scientists huddled together in the back of the laboratory, conspiring to ensure that evolution continued its reign of unchallenged deception.[14]

Don't be so quick to believe YEC scientific conspiracy theories. Also remember our example of the brain surgeon. Just because some YECs are committed Christians who love the Lord doesn't make them authorities on matters of science. In fact, very few YECs even have any professional scientific training or experience.[15]

Old-Earth Creationism has no problem accepting the geophysical data indicating the age of the earth, but again insists that the Bible supports an old earth. This also leads to very strange interpretations of Scripture that would have made no sense to the original audience. For instance, the many verses that refer to God "laying the earth's foundations"[16] of the earth are supposed to be describing how radiometric isotopes were miraculously placed into the earth's crust in quantities sufficient to enable the building of continents.[17] Again, it is simply unrealistic to expect the biblical authors to have possessed advanced knowledge of these modern scientific concepts.

But What about Evolution?

I wish the book could end right here. I wish I could just say, "Here is a better way to read the Bible that doesn't require us to mine God's Word for scientific details that can be applied to the universe as we know it today"— end of book. But if I did that, every reader would be left with the same question: this all sounds great, but how does it apply to Darwin's theory of evolu-

---

[14] Karl Gibberson, *Worlds Apart: The Unholy War Between Religion and Science* (Kansas City, KS; Beacon Hill Press, 1993), pg. 173.

[15] For a listing of the questionable credentials of the leading Young-Earth Creationists, see Volume 9, no 6 (1989) of the National Center for Science Education: Reports, 15-16, printed by the National Center for Science Education. You can see a more up-to-date listing on http://www.talkorigins.org/faqs/credentials.html.

[16] Isaiah 48:13; Zechariah 12:1, for example

[17] Hugh Ross, *The Creator and the Cosmos* (Colorado Springs, CO; Navpress, 1993), pg. 25. This same argument can be found at http://www.reasons.org/resources/fff/2000issue03/index.shtml #big_ bang_the_bible_taught_it_first.

tion? After all, there's nothing too offensive about reconstructing cosmic history from the clues that nature leaves us; most Christians can live with that. The Christians I discuss this with actually feel quite liberated after understanding the ancient Near-Eastern cosmological context of Genesis. In fact, we can avoid many unnecessary conflicts between science and religion simply by adjusting our expectations of both natural and special revelation—without ever having to extend this logic to the issue of biological origins. Why ruin it by talking about evolution?

Let me put everybody at ease by first saying that no Christian has to adopt an evolutionary framework in order to make sense of natural history. With God at the helm of creation directing the course of nature, Christians have an unlimited supply of miracles that can be dispensed at will to explain just about anything that science can't. It's like playing Monopoly with an unlimited stack of "get out of jail free" cards. And if it makes us feel more spiritual to get ourselves out of jail without having to roll the dice over and over again, then science can't stop us.

But if the same laws of nature that God used to bring order and structure out of the cosmic chaos were also used to create life on earth, then certainly there should be some evidence of this. Should we be afraid to at least look for it? And when all of the facts are laid out, they should start to tell a consistent story. Sure—some of the subplots and characters might still be unknown, and a few things may not make perfect sense, but what does the overall story indicate? What are the consequences if the story is *authentic*? What are the consequences if the story is *apparent*? How many times can we use our "get out of jail free" cards before the story of creation becomes a theological absurdity?

PART IV

WHAT ABOUT EVOLUTION?

CHAPTER NINE

A RECORD OF CREATION

As much as I would like to avoid it, there is absolutely no way to have a serious discussion about science and religion without addressing the spiritual challenge of evolutionary biology. I'll be the first to admit that I'm not too crazy about any suggestion that mankind has descended from something less than human. In fact, I personally don't like the idea at all. On the surface, it is offensive, disgusting, and seems absolutely contrary to everything we believe about God and ourselves. The whole thing just seems downright creepy to me. So trust me when I say that I would rather just gloss right over the entire topic and pretend it wasn't even a problem. But for many Christians, it is a problem—a *big* problem.

Evolution is opposed by most conservative Christians primarily because it is seen as contrary to the teaching of Scripture. If the Bible is clearly opposed to the idea of evolution, then for many of us that should settle it—end of discussion. But if Christians were to approach the creation narrative as a divinely inspired cosmogony that uses primitive non-scientific language to convey timeless truths about humanity, God, and nature irrespective of what each generation's contemporary scientific theories might be, then perhaps there are other clues by which we can piece to together the story of our natural origins. In other words, what if a serious investigation into the discernable patterns of providence by which God governs the universe is able to tell us something about the natural history of life on Earth?

For Christians, there are really only two ways that God could have created all living things: (1) by the continuous operation of the laws of nature, or (2) by a series of miracles that disrupted the regular patterns of material behavior. Neither of these creation methods is inherently more "spiritual" than the other. In fact, we can see examples in the Bible of God using both natural and supernatural processes to accomplish His will. Some Christians might argue that being a product of natural processes is not befitting of man's special status. If that is true, then I guess Adam and Eve were the only "special" humans since the rest of us obviously came about by "natural" processes. After all, whether one agrees with evolution or not, we can all acknowledge that our physical bodies were naturally assembled from pieces and parts of other organisms (i.e. the plants and animals that we eat), and that somehow this new material arrangement bears the image of God.[1] Of course, no Christian would argue that all men were not created in the image of God just because every human since Adam and Eve was born of flesh. But philosophically, the process of *evolution* is no different than the process of *gestation*, when applied to our genetic code over millions of generations. This just reinforces the point that "creation" can take place naturally or supernaturally. Ultimately, if God's will is to have creatures on Earth that bear His image, then either way should be fine with us—let God be God.

Many readers will understandably be skeptical. Why are we even considering this garbage? What advantage could there possibly be for Christians to embrace a Darwinian approach to natural history? In a debate that has evolved into an epic "culture war" between biblical theism and atheistic materialism, why would Christians want to cede any ground to the enemy? All of this sounds so radical, so novel, so liberal, and so—dare we say—*heretical*. Would we be the first generation of Christians to believe this way? Is there any historical precedent for such an approach?

Ironically, the idea of evolution was not perceived as much of a threat to religion when it was first proposed as it is today. Historic Christian orthodoxy has always maintained that God is both *transcendent* (separate from creation) and *immanent* (intimately involved with creation). But transcendence without immanence is simply deism, and immanence without transcendence is nothing more than pantheism. A biblical view of God and creation requires us to hold these two ideas in tension.

Many of Darwin's orthodox Christian contemporaries were actually quite comfortable with his explanation of the fossil record serving as the secondary causes behind God's governance of Earth's biosphere over time. Rather than isolate the creative powers of God to a single event in the distant past (transcendence), evolution seemed to emphasize God's continual creative providence over nature (immanence). The Reformed theologian James Orr (1844–1913) said the following:

> Assume God—as many devout [Christian] evolutionists do—to be immanent in the evolutionary process, and His intelligence and purpose to be expressed in it; then evolution, so far from conflicting with theism, may become a new and heightened form of the theistic argument. The real impelling force of evolution is now from within...[2]

Prominent Presbyterian theologian B. B. Warfield (1851–1921) took a notable interest in the scientific developments of the 19th century and had this to say:

---

[1] While in the womb, we are assembled from the plants and animals consumed by our mothers.

[2] Quoted in David N. Livingstone, *Darwin's Forgotten Defenders: The Encounter between Evangelical Theology and Evolutionary Thought* (Grand Rapids, MI: Eerdmans, 1967) pg. 142.

I am free to say, for myself that I do not think there is any general statement in the Bible, or any part of the account of creation, either as given in Gen. I & II, or elsewhere alluded to, that need be opposed to evolution.[3]

These are just two examples, but there were many others who had similar thoughts and feelings.[4]

Over time, the Christian position was taken over by "fundamentalist" believers and the theory of evolution was mostly defended by atheists and deists who require a godless worldview. As a result, evolution became inseparably linked to anti-Christian philosophies. This left most Bible-believing evangelicals with no choice but to oppose evolutionary biology as a morally bankrupt alternative to the biblical creation story. The barrage of religious attacks that followed was successful in turning evolution from a possible mechanism by which God intimately governs the course of biological history and gloriously demonstrates His continuing creative powers, into a repulsive and godless idea completely contrary to orthodox Christian belief. Charles Darwin has literally been blamed for such things as racism, genocide, rape, homosexuality, crime, and abortion.[5] As a result, even folks with no particular religious commitment are now prone to take offense at an evolutionary approach to the biological sciences.

All of these emotions are natural and understandable, but what does responsible science demand of us? Do our personal feelings have any right to inform us about what is scientifically true or false? Some medieval Christians felt completely justified in rejecting the Copernican solar system because of a misguided belief that it removed mankind as the center of God's attention and elevated hell, which was believed to occupy the center of the earth, up into the heavens—upsetting the whole plan of salvation. Should our own ideas of self-importance dictate what the spatial relationship between heaven and earth should be? Do we allow for that kind of personal bias to determine what we should believe about the natural world?

If the universe was in fact given the material properties to produce and sustain life through the discernable patterns of divine providence, and these laws of nature have been in continuous operation since the formation of the earth, then there should be some evidence of a gradual development of fauna and flora over time.[6] In fact, science should allow us to put the biological puz-

---

[3] Ibid, pg. 118.

[4] Ibid.

[5] As far as I can tell, these things were all around long before Darwin's time.

[6] Fauna and flora are just fancy ways of saying animals and plants.

zle pieces together and get a pretty coherent picture of how all of these things came about. Does this sound familiar? For those readers who have never heard of this, it is simply the science of evolutionary biology.

I know I've already said this, but I feel obligated to remind readers that my point here is not to argue for or against any specific scientific theory. A non-scientist like me has no business publishing definitive works for or against them. My intent is for Christians to properly understand *how* the scientific method is used to tentatively interpret the clues that nature leaves us. And when it comes to the history of life on Earth, nature has left us many clues. Of course, any explanation of the data that makes perfect sense today could be completely overturned tomorrow or a hundred years from now by new evidence, and it's not my job to keep up with the latest developments.

History shows us that scientific theories come and go as the process of discovery continually sheds new light on existing ideas. The point of this part of the book is to take an honest and objective look at the evidence of biological origins as it exists today, what we do know and what we don't know, and try to make sense of it in a way that still honors God as the Creator and Sustainer of all things.

Interpreting the Facts of Natural History

Unlike the sciences of cosmology and geology, evolutionary biology carries a lot of philosophical baggage with it. The continuous operation of the laws of nature may be an acceptable way to characterize the first 10 billion years of creation, but many of us would prefer for God to work *outside* of the laws of nature when it comes to life on Earth. On the surface, this seems like a reasonable compromise with science. After all, the consequences of the Big Bang and cosmic evolution are much "easier to swallow" than the consequences of evolution. As a result, many folks, even some professional scientists, are more prone to doubt a natural explanation of earth's biosphere than they are to doubt a natural explanation of the inanimate universe. Combine this personal incredulity with the anti-Christian materialistic philosophy that often accompanies evolution and many rational believers must "draw the line" at the feet of Darwin. As a result, it is often difficult for Christians to even understand evolution, much less have an objective unemotional opinion on it. So before we look at the evidence together, let's quickly review some basics.

Is evolution a fact? Facts are simply little pieces of knowledge about the world. All facts must be interpreted before they can mean anything. There is no such thing as a "pure fact" that just floats out in the marketplace of ideas all by its lonesome. In order to interpret facts, one must have an *interpretive framework* that allows the facts to mean something. Recall our "snow" example from Chapter One. The facts told us that it was the dead of winter, freezing cold, and the neighborhood awoke to 18" of fresh powder with no vehicular tracks. Our interpretive framework led us to the conclusion that it must have snowed overnight. Our neighbor, on the other hand, started out with a different interpretive framework. Rather than explaining the facts in terms of nature, he insisted that a fleet of levitating snow machines delivered the snow artificially. The same facts we used to conclude that it snowed were forced to fit into his framework by making different assumptions.

In both cases, the facts were made to fit into a pre-existing interpretive framework. And each framework attempted to explain the evidence as consistently as possible. If both approaches can sufficiently account for the data, how do we decide which interpretive framework is correct? In our example, common sense and Occam's razor clearly supported the more simple and straightforward approach. The "snow machine" hypothesis just seemed too forced and contrived and made too many unnecessary assumptions. We can use these same tools when looking at the facts of natural history.

Another way to decide between two interpretive frameworks is to push each one to its logical conclusion and examine the necessary consequences. Then the question must then be asked, "Can we live with these consequences?" In the case of our snowy morning, there is little consequence to the "it snowed last night" hypothesis. It's something that we should expect during the winter. But the "snow machine" hypothesis raises many more questions than it answers. What kind of person commands a fleet of levitating snow machines? What would be the motive for deceiving rational people? Why would somebody go to such great lengths to artificially replicate an event that seems to have a natural explanation? What is this world coming to?

This type of analysis can be very effective, but it is not absolute. If anything, our simple example shows us that facts can be forced into just about any interpretive framework without necessarily making that framework true. But as long as the facts keep fitting in a coherent and reasonable way, we keep using the framework. If the framework fails to account for the data or if the facts require too much awkward forcing or too many wild assumptions just to fit, then perhaps the framework needs to be adjusted or scrapped altogether.

In science, we call these interpretive frameworks *theories*. Evolution is the theory that most professional scientists use to interpret the facts of natural history. As with all scientific theories, evolution does not allow for supernatural causality, but is only able to consider material cause and effect. Understandably, this makes many Christians skeptical. We want science to confirm the supernatural realities that we all know to be true—at least when it comes to creation. For some reason, Christians are not that concerned with methodological naturalism when talking about "germ theory" or "quantum theory" or any other scientific discipline. We don't look to science to detect the providence of God as it relates to the weather or to planetary motion; but when it comes to the issue of our biological origins, naturalistic theories are usually ruled out a priori by Christians.

I don't think this is a theological necessity, but given the love affair that atheists and agnostics have with the theory of evolution, I can't really blame my fellow believers for taking this approach. It certainly doesn't advance the cause of science when the enemies of Christianity seem hell-bent on using evolution to beat our faith out of us. But as long as we agree with *them* that the claims of evolution are in direct competition with the Bible to explain the origin of life on Earth, we should expect this behavior. What better way for the world to discredit the Christian religion than with an evolutionary approach to natural history that excludes all supernatural influence? If evolution can explain everything, then what is left for God to do?

To counter the inherent naturalism of the evolutionary framework, many Christians have adopted interpretative frameworks that include supernatural causality. This makes most scientists skeptical. There is a very longstanding tradition of methodological naturalism in the sciences. As a procedure governed by rules, science is typically only able to explain material cause and effect by the continuous operation of the laws of nature. I've consistently argued that stretching science beyond its naturalistic limits in an attempt to provide a scientific basis for claims of faith only demotes the timeless transcendent truths of the Christian religion to tentative scientific theories and subjects them to falsification—and I stand by that. But consider this twist: if God were to work outside of His own system by performing a miracle, shouldn't the physical discontinuity caused by any divine disruption in the laws of nature be detectable?

The Intelligent Design movement makes a very interesting point here. They assume that any miraculous action taken by an intelligent being would have real effects that could be observed, measured, quantified, and documented. Since these effects would have been directly caused by divine action, there is not likely to be any physical evidence of secondary (i.e. natural) causes. In principle, such an investigation seems to fall within the scope of methodological naturalism. The only rational conclusion is that any such "causeless" effects must be evidence of an intelligent being who occasionally works outside the laws of nature. There is still the question of whether or not this conclusion can be considered *scientific* by definition, but it is certainly a logical *philosophical* conclusion.

For example: we go to a wedding with Jesus. The jugs are full of water one minute and then they're full of wine the next minute. The wine-water transition is fully supported by the facts, but since there is no known material mechanism that can cause this instant transubstantiation, it is clear evidence of a *discontinuity* in the laws of nature. The only rational conclusion is that this discontinuity (or singularity for you physics types) was caused by a miracle. Obviously, having Jesus take direct responsibility for the event in question makes the scientific investigation unnecessary. But since the world doesn't care what Jesus has to say about anything, we are forced to investigate these candidate miracles with the tools of science in order to prove our point.

If we believe in God, we believe in miracles, and all truth belongs to God, why not allow the scientific method to reveal the fingerprints of God in creation? Obviously science has no ability to speculate on the identity of the supernatural agent, but if it can simply pinpoint the absence of a material cause, why not allow it to do so? If a singularity is found, then the scientists can hand the investigation off to the philosophers and return to their laboratories to work on something else. It almost seems irrational for science to exclude the search for supernatural causes! Is science up to this task?

This interpretive framework can be applied to any investigation of natural history. An example from cosmology might be the Big Bang. Since the event is believed to have proceeded from a physical singularity—marking a discontinuity of the known laws of physics—the actual "cause" of the Big Bang is physically undetectable. Whatever "caused" it was clearly outside of space and time. Is this the fingerprint of God? Can such speculation be scientific, or should we move this discussion from the *science* classroom to the *philosophy* and *religion* classrooms?

Another example from cosmology might be the alleged billion-to-one accounting error during the period of matter/antimatter creation and annihilation. Is this event merely unexplained, or is it completely unexplainable? If it is truly unexplainable, then is that also the fingerprint of God? What if we say that it is a miracle, and then physics comes up with the scientific answer? Have we just given science another straw-man to further marginalize the Christian faith? Scientists who support the Intelligent Design (ID) movement can find many such events that appear to defy natural causes and effect. Each of these candidate miracles could be proof that God is intimately involved with His creation, or they could just be proof of scientific ignorance. The challenge for ID theory is being able to distinguish between the *unknown* and the *unknowable*—another tall order for Creation Science.

Despite the strong tradition of methodological naturalism in the physical sciences, ID arguments tend to resonate well with people of faith. And why shouldn't they? We know that somehow God is responsible for everything we see in nature. His handiwork is clearly evident throughout creation. We also know that there are many things that have not been explained in terms of science. When you put this all together, it seems completely rational to assume that when the scientific trail runs cold and no material explanation seems possible, we have just found unmistakable evidence of God working "outside of the system" to achieve His divine purposes.[7]

By using a theistic framework that allows supernatural causality, such as ID theory, Christians can interpret the same data used by evolutionists in a way that is consistent with our understanding of the Bible. On the surface, who can argue with this? But before we examine the logical consequences of each interpretive framework, let us first walk outside and see how much snow is on the ground.

The Fossil Record

For the majority of human history, the universe was seen as fundamentally unchanged since the time of creation, or at least since the Noahic flood. The seasons came and went, generation after generation of plants, animals and people lived and died—each descending after their own kind. The sun rose and set each day. Tides ebbed and flowed. The phases of the moon waxed and waned. Eclipses and comets could be predicted like clockwork. There was really no reason to believe that life on planet Earth was ever any different than the way we had known it to be throughout the course of human history.

All of that changed during the Industrial Revolution of the late 18th century. Roads and canals were being carved through mountains to open lanes of shipping and transportation. These massive earth-moving projects revealed a version of the past that was much different than anyone had previously imagined. Once people started looking carefully at the layers of earth laid down over the ages, they came to the startling realization that life on planet Earth did not always look like it does today.

These excavations literally uncovered hundreds of thousands of *fossils*, the mineral remains of dead creatures that once populated the earth. In fact, it looked as though many thousands of generations of unknown plants and animals appeared and disappeared, all the way up the geological column. Each successive layer of earth revealed subtle differences over time. It seemed as though no matter where we looked, the earth was telling the same story, the story of *faunal succession.*[8]

Faunal succession tells us that no matter where in the world people dig, certain layers of earth always contain certain types of plant and animal fossils

---

[7] Of course, Paul tells us in Romans that all of creation testifies to the handiwork of God. So do we really need a scientific investigation to detect it? And if we hand science the power to "find" God, haven't we also just given them the power to dismiss God? These are some of the philosophical issues that need to be addressed by the ID community.

[8] This discovery preceded Darwin's theory of evolution by 50 years or more.

unique to that layer, and each layer is slightly different than the layer before it and the layer after it. These layers are also known as the *fossil record*, and they allow scientists to reconstruct the biological history of the earth. Consequently, looking deep into the earth's crust is very similar to looking deep into space. It is a crude "time machine" that allows *paleontologists*[9] to see how life on Earth has changed through the ages.

Using the same principles of radioactive decay that we looked at in Chapter Eight, each layer of earth can be dated. And when the geologists and the paleontologists get together and lay everything out from youngest to oldest, some very obvious and undeniable patterns begin to emerge. For instance, single-celled bacteria, the simplest and arguably the heartiest of all the living creatures were the first on the scene. Photosynthetic bacteria and algae were the next things to appear. Nothing but the remnants of bacteria are ever found in layers of earth older than 1.5 billion years. Multicellular organisms begin to appear in layers of earth that are roughly 700 million years old. The layers of earth between 500 and 570 million years old are marked by an explosion of multi-cellular life such as shellfish and corals. These primitive organisms were followed by fish and seedless terrestrial plants, then amphibians and insects, seed bearing plants, reptiles, dinosaurs, small mammals, birds, flowering plants, the extinction of dinosaurs and finally large mammals.

One thing that is important to note is the unmistakable consistency of this basic sequence. Reptiles are never found before the first appearance of amphibians, amphibians are never found below the first appearance of fish, flowering plants are never found below the first appearance of seed-bearing plants, which are never found below the first appearance of seedless plants. In fact, this pattern is so universal that it is often referred to as the *law* of faunal succession. Moreover, the fossils found in a geographic location appear to be the ancestors of the living species at that location. In other words, the fossils found in Australia are anatomically similar to the unique variety of plants and animals currently living in Australia, and are not found anywhere else.[10] The fossils found in South America resemble the living species unique to that area. This same pattern of biological change within a geographic region holds true for North America, Africa, the Galapagos Islands and just about every-

---

[9] A paleontologist is someone who uses fossils to piece together the natural history of the earth's biosphere.

[10] Different mechanisms of species dispersal, such as plate tectonics and migration, must also be considered when making sense of the data.

[11] Interestingly, the plants and animals found on the Galapagos Islands appear to be most closely related to found in South America, consistent with the prediction that the Galapagos were once populated with various plants and animals from South America, their nearest geographic neighbor.

where else that things lived and died.[11]

Another important thing to note is that whenever a new major group emerges, the first members of that group have many characteristics of the preceding group. For example, the first amphibians to appear in the fossil record have more fish-like features than modern amphibians do, such as internal gills and fish-like forelimbs. The first reptiles to appear in the fossil record have amphibian-like features, and the first mammals have reptile-like features.

What does all of this mean? What are we to make of this? I'll sum this up with the following quotation from a Christian professor of biology:

> Time after time, species after species, the greater our knowledge of the earth's natural history, the greater the number of examples in which the appearance of a new species can be linked directly to a similar species preceding it in time. These histories reveal a pattern of change, a pattern that Darwin aptly called not "evolution," but "descent with modification." Once this pattern becomes clear, and it can be found in any part of the fossil record, the theme of life is equally clear.[12]

And this "theme of life" is not just random incoherent change, but a logical succession of biological diversity from the simplest single-celled bacteria to intelligent humans that have the cognitive abilities to figure all of this stuff out simply by digging through the earth's crust.

Already we have a serious problem for those who want to understand the Genesis creation narrative as a scientific account for life on Earth. The actual fossil record is quite different than what one would expect from a relatively recent creation event in the vicinity of modern Iraq over a period of six days. Life seems to have *appeared* and *disappeared* everywhere in stages, over billions of years, with each succeeding stage slightly different than the previous one.

Some in the creation science movement have tried to explain the fossil record in terms of the great flood of Noah's time. As the flood waters rose, the more primitive organisms drowned first while the more complex species ran to higher ground and were buried last, thus producing the observed pattern of faunal succession. But when radiometric dating is applied to the layers of sediment supposedly deposited during the great flood, they are found to be separated by hundreds of millions of years. Moreover, the fossil record clearly

---

[12] Kenneth R. Miller, *Finding Darwin's God: A Scientist's Search for Common Ground Between God and Evolution* (New York, NY; Harper, 1999), pg. 40.

shows that many thousands of plant and animal species became extinct in those layers. Does that mean Noah only gathered a tiny fraction of the earth's animals? In addition to understanding how a massive and violent flood could have perfectly sorted all animal remains into such a consistent pattern, it's also difficult for some people to visualize seeded plants "outrunning" non-seeded plants, and flowering plants "outrunning" non-flowering plants as the flood waters rose. Sometimes faith can only go so far. At some point, we should return to the Bible and see if perhaps we've missed something in our understanding. Thankfully, many evangelical Christians are starting to do this.

As amazing as the fossil record is, it will probably always be incomplete. It is very unlikely that each and every step of evolutionary change will ever be fully documented with fossil evidence. I hate to keep using the analogy of a puzzle, but it makes a lot of sense. Although scientists can see the overall picture of how life on Earth has changed over time, there are still many specific pieces of the puzzle that are missing. In many cases, there may be no fossil evidence of a transitional species that links two known groups together, leaving scientists to merely speculate on what these creatures may have looked like.

Some periods of natural history also seemed to have evolved at a relatively slow rate before undergoing short periods of rapid[13] change as new species seem to have appeared "out of nowhere." It's quite possible that the scientists may never find these missing links. But the fact that thousands of fossilized transitional forms have already been found, including some of the more challenging examples like the link between land mammals and sea mammals, gives scientists hope for more future discovery.[14]

One thing that the fossil record does not do is provide us with a material *mechanism* of evolutionary change. Finding fossil evidence clearly showing that the first amphibians had many fish-like features doesn't really tell us *how* a population of fish might have evolved into a population of amphibians over a few hundred million years. And we'll never be able to directly observe a population of fish evolving into a population of amphibians. Likewise finding the fossilized remains of a sea-going mammal that can also walk on land does not tell us *how* a population of mammals could have returned to life in the ocean. Unfortunately, nobody will ever have an opportunity to observe that happen either.

For those scientists who take an evolutionary interpretation of the fossil

---

[13] *Rapid* is used as a relative term. Usually it means that after tens of millions of years of *slow* change, we find several million years of *rapid* change. Even these periods of so called "rapid" change would be about as exciting as watching your grass grow.

[14] There are numerous transitional forms now known between Artiodactyls (cows, hippos, camels, rhinos, etc…) and whales, including *Ambulocetus*. Hippos appear to be the closest relatives of whales, though cows are in the same general group.

record, the most popular theory of what directs evolutionary changes is called *natural selection*. Each species has a certain amount of variation caused by random changes (mutations) in their genetic code. These genetic changes can be beneficial, like the changes that allow insects to defeat insecticides and bacteria to overcome antibiotics. Or they can be detrimental, such as birth defects and congenital diseases. The environment selects which changes get passed on and which changes reach a dead end. The more beneficial a genetic change is, the more likely it is to be inherited by the next generation. The theory of evolution claims that over 3.8 billion years, these tiny steps have produced all living things from the first single-celled organisms.

Whether the fossil record happened by many miracles (special creation) or through the continuous operation of the laws of nature (evolution), the one thing that is evident from the pattern of faunal succession is that life has changed drastically over time.

Applying an Interpretive Framework

There is absolutely no way I'm going to prove to anybody that the evolutionary interpretation of the fossil record is or isn't the correct approach. Moreover, there are many different ways to interpret the fossil record supernaturally. Darwinists, YECs, OECs, and ID theorists go around and around in circles discussing this kind of stuff. Rather than get swept up in that mess, I'd like to take a slightly different approach. Let us suppose, for the sake of argument, that evolution is *impossible*. Let us start with the assumption that a new species simply can't emerge no matter how much time evolution is given. Despite the Creator's meticulous fine-tuning of the physical cosmos to support life on planet Earth, God either can't get it done, or doesn't want to do it, by working within His own system. If anything biological is ever going to happen in this universe, God must suspend His own patterns of providential governance and perform a miracle or series of miracles. But unlike the biblical miracles that were performed specifically to draw attention to the Sovereign Lord of heaven and earth, the "miracles of creation" are strangely subtle and well hidden—almost indistinguishable from the ordinary patterns of providence. Is this presupposition even necessary? Do Christians stand to gain anything from making this assumption? Do we stand to lose anything? I ask you to consider these questions as we go through this little exercise.

With this precondition firmly established, let us take a look at the fossil record and try to make some sense of it. When it comes to evolution, the fossil record is like the 800 pound gorilla that many Christians don't want to acknowledge. Young-Earth Creationists don't even take it seriously, or summarily dismiss it all as a relic of Noah's great flood. Old-Earth Creationists fully acknowledge all 3.8 billion years of the fossil record, but tend to downplay the transitional fossils and instead focus on the "sudden appearance" of many species as proof that evolution is impossible. This probably isn't a good long-term strategy. What happens when those gaps are eventually filled in with newly discovered fossils? In fact, whenever a transitional fossil is found, it is found exactly where it should be in the pattern of faunal succession. If any transitional species were ever found in a layer of strata beneath that of their ancestors, it would be a serious challenge to the evolutionary framework. But no such thing has ever been observed. The discovery of transitional fossils in their proper geological layers is now so commonplace that it rarely gets any attention outside of the scientific community.

Attacking the fossil record for its incompleteness may have been an effective strategy against evolution 100 years ago, but so many things have been found since the days of Darwin that most well informed opponents, such as some in the ID movement, have given up those tactics.[15] It's true that the fossil record is, and probably always will be, woefully incomplete. But even the missing data have to be interpreted by a framework.

If we take an evolutionary interpretation, these gaps are just the result of incomplete knowledge. The fossil record is therefore not really a *detailed* record of the history of life on Earth, but rather a *partial* record of those creatures that happened to die under just the right circumstances to be preserved. It's quite possible that the vast majority of Earth's plants and animals became extinct without ever having met the specific conditions for fossilization. Millions of species that lived and died might never be discovered. That seems reasonable.

On the other hand, if we take a supernatural interpretation, then these challenging gaps might suggest that many species appeared suddenly—ex nihilo—without any evolutionary precursors. They also provide good ammunition for anti-evolutionists to challenge the possibility of a species evolving into another species naturally. Certainly any scientist is free to challenge the existing evolutionary paradigms, but I have an important suggestion that Christians everywhere need to consider: rather than attacking the *missing* pieces of the evolutionary puzzle like over-zealous defense lawyers trying to gum-up the prosecution's case, why don't we face off with the 800 pound gorilla—seriously consider the evidence that *has been found* and carefully think about how to deal with the spiritual implications of that first? Forget about the

---

[15] Michael J. Behe, one of the leaders in the Intelligent Design movement wrote: "For the record, I have no reason to doubt that the universe is the billions of years old that physicists say it is. Further, I find the idea of common descent (that all organisms share a common ancestor) fairly convincing, and have no particular reason to doubt it."—From his book: *Darwin's Black Box: The Biochemical Challenge to Evolution* (New York, NY; Free Press, 1996), pg. 5.

so called "missing links" for a minute. Given what we already know about Earth's natural history from the fossil record, there are some pretty big questions that believers need to answer before attacking the apparent gaps.

Don't just read through this list, but really ponder these questions carefully. Think about how you would answer them, and what those answers would say about God. For instance, if evolution is fundamentally impossible, then:

- Is the appearance of each new species in the fossil record a separate act of creation by divine fiat?

- Would God have allowed some species to evolve up to a point, only to wipe them out and replace them with a similar one in the same geographic region but upgraded with a few new features?

- Did the extinct species found in the fossil record become extinct *naturally* or did God use a miracle to wipe them out and start over with slightly upgraded versions?

- If they did become extinct naturally, was this a creative failure on God's part? Did God have to keep making new species because none of His 'designs' were 'intelligent' enough to stand the test of time?

- Did God keep repeating this process of creation by "trial-and-error" until He finally exterminated the *Cro-Magnons* and the *Neanderthal* only to replace them with Adam and Eve?

- If He did not, then were these early hominids[16] related to Adam and Eve?

- Was the entire fossil record just put in the ground by God to maintain a consistent illusion of natural history?

If we assume that evolution is impossible, then the creation of a new living species can only happen by a supernatural act. No Christian would deny that this is possible since God can do anything He wants by performing miracles. But the question is not *could* He have done it this way, but did He *have* to do it this way? In other words, were miracles God's only creative option? And if so, why did He try to "hide His hand" by making these miracles look like they could have happened naturally, through a gradual process of change over long periods of time?

If we take the "many miracles" approach, then we must interpret the fos-

---

[16] The hominids are the "great ape" family (Hominidae)—which includes humans, chimpanzees, gorillas and orangutans.

sil record as the history of supernatural creation—a literal index of what God made, when He made it, and where He made it. This puts God in the awkward position of having created all living species, even humans, by a process of "trial-and-error" over a period of 3.8 billion years. But rather than give us unmistakable evidence of special creation by mixing up His creations and distributing them evenly throughout the entire fossil record, God instead creates everything according to a very specific geological and geographical distribution that looks very, well—natural.

For example, single-celled bacteria dominated most of Earth's natural history. Why did God play around with bacteria so long before moving on to multi-cellular life? Did it take Him that long to figure out how to do it? Moreover, the vast majority of God's creations, including our closest known hominid relatives, didn't even survive the test of time. Were these extinctions done on purpose, or were they failed prototypes—part of God's ongoing "research and development" program?

These are the tough questions raised by the fossil record. How are we supposed to answer them? Does interpreting the fossil record using an evolutionary framework raise these kinds of difficult theological questions? To help sort this out, we can compare the conclusions drawn from each system. For example, evolution teaches us that it took a billion years for these bacteria to pump enough oxygen into the earth's atmosphere to sustain more complicated forms of life.[17] That seems like a reasonable scientific conclusion. Ironically, some creationists also claim that God created these bacteria first just to help prepare the earth for more complicated forms of life.[18] For instance, these early bacteria converted poisons into metal ores, removed methane from the primordial atmosphere, and reduced noxious sulfates.

This all sounds pretty neat, but why does God need to miraculously create bacteria to prepare the earth naturally, just so He can then create multi-cellular life with another miracle? How exactly does the Maker of heaven and earth find Himself at the mercy of single-celled bacteria? If He is going to ultimately "cheat the system" by creating both bacteria and multi-cellular life using miracles, why not just "wave the magic wand" one more time and make the earth ready for multi-cellular life a billion years sooner? Or why not just skip primitive multi-cellular life altogether and miraculously create the entire biosphere as we know it today in its fully developed state? Does this interpretation not seem slightly awkward?

To believe that bacteria *had* to exist all by themselves for over a billion years to prepare the earth for multi-cellular life assumes that the laws of na-

---

[17] Jon Copley, "The Story of O," *Nature* 410 (2001): pp. 862-864.
[18] Hugh Ross, "Bacteria Help Prepare Earth for Life," *Connections 3*, no. 1 (2001), pg. 4.

ture were acting continuously throughout the earth's natural history. And if the laws of nature were allowed to run their course, why was God left with no other option but a miracle to finish the job? If God indeed created the heavens and the earth from nothing, sustains their very existence by the Word of His Power, and governs every particle of matter according to His perfectly designed laws, why would He be forced to cheat His own system in order to close the deal? This seems very inconsistent with God's sovereignty, omniscience, and omnipotence.

Even after multi-cellular life finally appears on the scene, there are other difficult questions. For instance, why did it take another 1.5 billion years for more complicated forms of life to really take off? And why did God create so many species only to make them extinct before finally creating man another 500 million years later? According to the theory of evolution, these many intermediate stages of life (along with the help of a fortuitous asteroid some 65 million years ago) were all necessary steps in a long chain of cause and effect that ultimately led to the emergence of modern humans about 200,000 years ago.

If we assume that evolution is impossible, then these data have to be interpreted differently. For example, one of the reasons given by a leading Old-Earth Creationist for why there was so much life and death before mankind is so that modern humans can enjoy an abundance of oil, which is made by decaying life-forms over millions of years.[19] So apparently, from the foundations of the world, God was quite concerned with man's insatiable appetite for fossil fuels. But rather than performing just *one* miracle to guarantee that mankind arrives on the scene with enough oil to last at least a few hundred years, God apparently performs billions of miracles over billions of years to create many diverse life forms, just so there is enough death and decay to give mankind a finite supply of oil. This all sounds terribly inefficient and wasteful, which ironically is a charge typically leveled by Christians toward any idea that God would create through an evolutionary process. If God is really that concerned with our use of fossil fuels, why not just miraculously give us an unlimited supply without all of the wasteful death and decay?

One of the toughest challenges presented by the fossil record is the geographic and geological pattern of faunal succession. The progression from bacteria, to multi-cellular life, to seedless plants, to fish, to amphibians, to seed-bearing plants, to reptiles, to mammals, to flowering plants, and finally to humans is universal. For any location in the world, each new set of species is anatomically similar to the preceding species. If we allow evolution to be

---

[19] Hugh Ross, "Petroleum: God's Well-Timed Gift to Mankind" www.reasons.org/resources/connectinos/200409_connectinos_q3/index.shtml

true, then the obvious explanation is that the pattern of faunal succession represents the history of descent with modification over 3.8 billion years. If we assume that evolution is false, then we are left with a strange pattern of special creation.

Apparently, God created each new species, slightly more adapted than the ones He previously wiped out, in the exact same geographical location as their apparent ancestors. Of course, nobody denies that God could have easily created life this way, but this kind of "trial-and-error" is usually how finite creatures with limited knowledge do things. Shouldn't God be greater than that? Rather than God being sovereign over creation and coherently working through the perfect laws of nature, the fossil record shows Him constantly tinkering with creation, often suspending His own rules to "cheat the system" so that a new species can "suddenly appear" despite having finely tuned the system precisely to sustain life in the first place! Again, does this not seem strange?

On the other hand, if we interpret the fossil record using an evolutionary framework (but not a materialistic worldview), then we simply conclude that God must have created life by the continuous operation of the laws of nature. What's so bad about that? The fossil record is then interpreted just as any other scientist would interpret it—as a consistent, but incomplete index of the gradual change of life over time. This certainly would spare Christians from a lot of unnecessary controversy. It also seems like a very sensible way to understand the obvious patterns found in the earth's crust and it takes nothing away from God as the ultimate Creator and Sustainer of the universe. Consider this: if the universe were a giant game of pool with 10^{80} balls, and the laws of physics were designed and operated by God alone, shouldn't He be able to sink every ball on the break without changing the rules during the game?[20] Even if He did "run the table" like a common pool shark, waiting to see the results of each divine shot before proceeding to His next move, it would still seem like He was less than God.[21]

As creepy as evolution first sounded to me, I'm actually starting to like it better than these other options. Not as an absolute philosophical description of the meaning and purpose of life—which is clearly outside the naturalistic boundaries of science—but as a tentative scientific framework of interpreting the fossil record. The "many miracles" approach, as a scientific explanation, just seems too awkward. Obviously God could have worked through a series of miracles over billions of years, but I just don't see any advantage to that

---

[20] Physicists calculate that there are roughly 10^{80} fundamental particles in the observable universe.

[21] I saw this example in Kenneth R. Miller, *Finding Darwin's God: A Scientist's Search for Common Ground Between God and Evolution* (New York, NY; Harper, 1999).

claim. And I don't think God would have put fossils in the ground just to fool us either. If the sovereignty of God over His creation can be expressed either naturally or supernaturally, why unnecessarily raise these other questions by invoking all of these miracles?

Theistic Evolution?

For some Christians, the evidence for evolution is as clear as the evidence that the earth revolves around the sun. We can't observe evolution directly and probably will never be able to, but like the concept of our solar system, the theory of evolution is clearly inferred from those things which can be easily seen, like the fossil record. Some readers may have already concluded that my position can best be described by the term *Theistic Evolution*, but that is not entirely correct.

Actually, I can't stand this term. For me, it ranks right up there with "divine intervention" as one of the most oxymoronic theological phrases ever put together. But I guess there is just too much materialistic philosophical baggage associated with unbridled Darwinian thinking for Christians to allow the term "evolution" to exist all by itself. What better way to sanctify an "inherently" atheistic belief than to slap the *theistic* qualifier in front of it? Rather than just having plain ol' atheistic evolution, we now have a "theistic" version just for Christians.

I'm sure that the term *Theistic Evolution* is meant to imply that God "guides" evolutionary change by His hand of providence, thereby separating the *science* of evolution from the materialistic *philosophy* that often accompanies it. But if we really understand the sovereignty of God, why would any natural process need such a qualifier? This only perpetuates the false notion that the continuous operation of natural forces apart from miracles necessarily implies deism or atheism. Do we do this with any other scientific discipline? Think about this for a minute.

Take the modern science of meteorology for example. The Bible clearly states that God "makes clouds rise from the ends of the earth. He sends lightning with the rain and brings out the wind from his storehouses."[22] Before mankind had the scientific tools to examine the patterns of natural behavior that cause changes in the weather, meteorological phenomena were directly attributed to divine providence without regard to secondary causes (the laws of nature). But the job of an atmospheric scientist is to try his or her best to explain the weather without reference to God. Are these meteorologists guilty of leaving God out of the picture? Does this make the naturalistic science of

---

[22] Jeremiah 10:13

meteorology a Godless and atheistic discipline? Of course not! Nobody cares about a meteorologist's theology when he's giving a forecast!

Be assured: God's sovereignty over the universe is in no way diminished by the implied naturalism of the meteorological sciences. The doctrine of primary and secondary causes reminds us that God's providential governance of meteorological phenomena is not in competition with the discernable patterns of material behavior that explain physical changes in the weather. And neither does this fact make the science of meteorology inherently atheistic. Understanding the weather in terms of the continuous operation of the laws of nature is a noble vocation, appreciated by both theists and non-theists alike. We don't have to stick the "theistic" qualifier in front of it just to prove that it's acceptable for Christians to believe in or practice religiously neutral meteorology.

Have you ever seen your weatherman acknowledge God's sovereignty over the atmosphere? I have not, but I can just see the "controversy" unfolding—Christians across the country calling their local news channels complaining about the "atheistic bias" propagated by network television and demanding that "theistic meteorology" be given equal time during the nightly weather forecast! Public school textbooks will have to be re-written so that the meteorological sciences rightfully acknowledge God's providential governance over the weather. Teach the controversy!

Of course all of this is just plain silly. But why then do some Christians have to qualify their belief in evolutionary biology this way? If a biblical theology of creation is indeed the basis for the uniformity of nature, then all of the natural sciences already assume that God is working through secondary causes to govern the internal affairs of the cosmos. So when confronted by fellow Christians for the unpardonable sin of believing in naturalistic science, why not instead take the opportunity to explain that of all people, Christians should not fear the outcome of honest scientific inquiry because all of the natural sciences are based on God's sovereignty over creation?

Another problem with Theistic Evolution is that it takes another tentative scientific theory and gives it the status of a religious doctrine. Have we not learned anything from history? What will happen to today's Theistic Evolutionists if common descent gets overturned by new data? If the theory of evolution is wrong, then let it be wrong by the testimony of natural history, not because some Christians mistakenly think that it is inherently deistic or atheistic. On the other hand, if evolution is not inherently deistic or atheistic, then can we please drop the "theistic" qualifier? This only makes Christians sound petty.

CHAPTER TEN

THE UNIVERSAL TREE OF LIFE

Like the Big Bang theory, the theory of evolution is widely misunderstood. For starters, evolution doesn't really concern itself with *how* or *why* life began. Scientists have successfully demonstrated that certain building blocks of life, like amino acids and peptide chains, can be created by simulating the natural conditions of a primordial tide pool, but arranging this material into something that can replicate itself is an enormous leap.[1] Whatever caused the chemical jump from non-life to life has eluded detection. Any actual evidence of this event would have been destroyed by the geological activity of the early Earth.

Scientists figure that whatever this event was, it preceded the oldest known samples of the earth's crust (3.8 billion years) because these earliest rocks contain evidence of already existing bacteria.[2] So unless self-replicating molecules can be created and observed in the laboratory, I doubt that anyone will ever know the actual material mechanism that organized these first "living" structures—provided there was one. Whether these first living things came to earth naturally or by a miracle, the science of evolution only cares about what happened over the next 3.8 billion years.[3]

Evolution in a Nutshell

The basic concept of biological evolution is really quite simple. Evolution is defined by biologists as an overall genetic change in a given population over time. The term *genetic change* implies that the new characteristics can be inherited by the descendants within that population. If I go around the yard and take a leg from every spider, the next generation of spiders in my yard will not be missing any legs. These kinds of changes don't affect the *ge-*

---

[1] The simulation I'm referring to is the Miller-Urey experiment from 1953.

[2] There is some debate as to whether the organic deposits found in these 3.8 billion year-old rocks prove the existence of life. The rocks are thoroughly metamorphosed but do contain curious bits of carbon with unusually stable isotope ratios. The chemistry of life is the most obvious cause of these ratios in carbon, but there is always the possibility that some other phenomena caused this.

[3] The scientific investigation into the actual "beginning of life" is referred to as *abiogenesis*.

netics of a population and therefore can't be passed down. Evolutionary changes are only those that can be inherited by the offspring in a population. The term *population* refers to a species, or an isolated group within a species that compete with each other for limited resources and mate with each other to produce offspring within the population. One can refer to a population of bacteria, plants, animals, insects or any other living species.

Another common misconception is that individual organisms can evolve by themselves. If you ever hear somebody confused over how a cow can turn into a whale, then they obviously don't understand evolution. In fact, if a cow did turn into a whale, or give birth to a whale, that would be very strong evidence *against* evolution. However, more subtle variations of this misunderstanding still persist. For instance, one might have the impression that giraffes "made their necks longer" by stretching up to reach the leaves of tall trees. This is false. An organism can't "will" itself to change its own genetics in response to the environment. Evolution is only concerned with heritable genetic changes *of a population* over time that occur when certain naturally occurring characteristics within a population are favored over others in response to the environment.

Evolution only works because every species has a certain amount of natural variation within a population. Consider a small hypothetical island populated with cats. Let's say you have some natural variation with respect to fur color. Our hypothetical cat population is 50% black, 25% white, and 25% orange. If someone were to revisit this "Cat Island" a few hundred years later and find the population to be only 10% black, 10% white and 80% orange, then they would conclude that the population of "Cat Island" evolved. In other words, something happened between then and now that favored the orange cats five to one over the other colored cats. I know this doesn't seem very dramatic, but that *is* evolution according to the simplest definition.[4]

Now some smart folks may be quick to point out that all the cats on Cat Island were all still *cats*. We didn't return to find *dogs* or *mice*. That is very true indeed. So what we witnessed in our hypothetical experiment is commonly called *microevolution*, or evolution within a species. Microevolution can be observed and demonstrated both in a laboratory and in the wild so it is generally accepted by both creationists and evolutionists.

What seems to give people heartburn is the idea of *macroevolution*, which is commonly understood to represent the evolution of one species to another. But according to evolutionary theory, many of these small micro-

---

[4] I don't know the first thing about feline genetics, so if that happens to be your area of expertise, I apologize. Hopefully other readers will appreciate this picture of changes within a population over time.

evolutionary changes can add up over millions of years and eventually a population will be different enough from its ancestors that biologists will classify it as a new species (speciation). So the term macroevolution is unnecessary and misleading since all evolutionary change is thought to proceed in microevolutionary steps—summed up over incredibly large spans of time. The main challenge with speciation is that we can't easily observe it (or define it), so we are often left to observe the smaller changes and speculate on how they might accumulate over millions of years to create a new species.[5]

When speciation occurs, the new species' ancestors, who may have migrated to a different environment, would have adapted specifically to their new environment, perhaps also evolving into a new species. When something like this happens, these two new species are said to have evolved from a *common ancestor*. That is, they both descended from the same population, but are now different enough from one another to each be classified as a new species.[6] In many cases, the common ancestor has already become extinct. As a result of being related to a common ancestor, these two new species will have very similar characteristics—both anatomical and genetic. These similarities and differences between species are how biologists classify all livings things.

The idea of common descent is also commonly misunderstood. For example, evolution doesn't say that man came from monkeys, or that whales came from cows, but that all primates descended from a common ancestor that was neither a human nor a monkey, and whales and cows descended from a common ancestor that was neither a whale nor a cow. Going back further, biologists can trace many of the species we know today backwards through the fossil record and sometimes with the help of genetics, find generation after generation of common ancestors all the way back to the oldest bacteria.

Does this sound just too far-fetched to be true? Is this just another sad case of the secular world grasping at straws in order to avoid any evidence of special creation? Obviously there are many unknowns when it comes to evolution. How exactly did cows and whales descend from a common ancestor? How did the individual components of complicated structures like the human eye all come together to form a fully functioning organ? What kind of natural processes could have possibly caused such improbable things? Despite the

---

[5] The kinds of microevolutionary changes that can be observed, both in the laboratory and in the wild, happen thousands of times faster than what is actually required to account for the pace of evolutionary change indicated by the fossil record, including the periods of so called "rapid speciation" predicted by the theory of punctuated equilibrium. These studies are cited in Kenneth R. Miller, *Finding Darwin's God: A Scientist's Search for Common Ground Between God and Evolution* (New York, NY; Harper, 1999), pp. 107-111.

[6] As far as I can tell, something is usually defined as a new species if it can no longer breed with the other descendants of the same common ancestor, but there is much more to it than that.

geological record showing that life has changed over time, for those of us who don't spend a lot of time studying the scientific details, the whole idea of evolution can stretch the imagination.

Regardless of exactly *how* evolution might have happened, good theories are able to make predictions that can be verified by observation. If the predictions are confirmed, then the theory hangs around. If the predictions are contradicted by the data, then we have to change the theory or toss it out. The theories of evolution and special creation are no different.

For example, if evolution is true, then all living things must have descended from a common ancestor in a "universal tree of life." As a result of this gradual development over time, the similarities and differences between all species should fit into a pattern of *groups within groups*, or *nested hierarchies*—clearly demonstrating their ancestral history. These patterns should be evident by anatomy as well as genetics. Moreover, for evolution to be a viable theory, this universal tree of life should be completely consistent with the geographical and geological distribution of all species found in the fossil record. If these predictions can't be confirmed, then the theory of evolution, as a framework of interpreting the data of natural history, is in big trouble!

Using our supernatural framework of special creation, we can also make some simple predictions about all living things. For instance, it's reasonable to conclude that the original ancestors of each species alive today were created ex nihilo by distinct acts of special creation, and each species descends only after its kind. If this is indeed the way it actually happened, then we should not expect the similarities and differences between species to follow any specific patterns. In the absence of such patterns, special creation would be confirmed as a reasonable framework for interpreting the data of natural history and the theory of evolution would be overturned once and for all.

Building a Family Tree

The pattern I'm talking about is one that resembles a family tree. For instance, if you took a set of grandparents, two sets of parents, and six children, you would have twelve members of a family tree. Now if you took five sets of these same three-generational family units, one African, one European, one Asian, one Native American, and one Scandinavian and mixed them all up, how hard would it be for someone else to rearrange them back into their family groups solely based on physical appearance? It probably wouldn't be that difficult since the family relationships would be obvious from external appearances.

This is similar to how biologists classify organisms. Only rather than using differences in appearance based on ethnic origin, they use anatomical characteristics based on biological origin. Grouping plants and animals together based on these physical similarities is called *comparative anatomy* and it doesn't just include what lives and moves on the earth today, but everything that is known from the fossil record as well.

For example, people are classified as *Homo sapiens*. We are the only species (*sapiens*) currently living of the human genus (*Homo*). Because of the numerous anatomical similarities between humans, chimpanzees and gorillas—like being tailless—we are all grouped into the same family (Hominidae). Going back further, the hominids all belong to the same order (Primates) as other creatures with similar features such as five fingers/toes, fingernails/toenails, binocular vision, and a similar dental pattern. The primates are in the same class (Mammalia) as dogs, cats, seals, rodents, marsupials, whales and dolphins because we are all warm-blooded, we nurse our young, and we don't lay eggs.[7] Mammals belong to the same phylum (Chordata—subphylum Vertebrata) as reptiles, birds, amphibians and fish because of our backbone and nervous system. And all chordates are obviously part of the animal kingdom (Animalia) along with non-vertebrates such as snails, clams, insects, spiders, jellyfish, worms, etc. The other kingdoms include plant (Plantae), fungus (Fungi), protozoa (Protista), and bacteria (Eubacteria/Archaebacteria), which have very few of the same anatomical characteristics as creatures in the animal kingdom.

When all living things are classified this way, a consistent pattern of nested hierarchies (groups within groups) begins to emerge. There is no logical reason, except for the theory of common descent that explains why all living things fit perfectly into a pattern of nested hierarchies. For example, if each species were actually created as separate and distinct, then we should expect to see a more even distribution of anatomical characteristics, as we see in other groups of things that were created separately. Take automobiles for instance.[8] When shopping for a new vehicle, you can mix and match just about any features that you want. You can get a truck or a car with either four or two doors, you can get a sedan or a coup with leather or upholstery, you can get a sports car or a station wagon with or without power windows or air, you can get two or four-wheel drive on a truck, a car, or an SUV.

The reason you can do this is because each type of vehicle was specifically designed and built as a separate and distinct model. They are obviously not in any way related to, or descended from one another. In other words,

---

[7] A few mammals lay eggs and are not as warm blooded as typical mammals.
[8] I credit this analogy to Douglas L. Theobald.

since automobiles do not fall into groups of nested hierarchies, the unique-ness of each model can be explained in terms of *design* plus *special creation*. The design of each vehicle is not constrained by family relationships.

If living things were similarly designed and assembled, we should also expect to see the designer mixing and matching features between various groups of organisms to maximize anatomical diversity. But when it comes to the actual "diversity" of life on our planet, this does not appear to be the case. Living things simply don't share their unique features with other creatures outside of their specific group, unless there is a common ancestor that also had that feature. No matter how fascinating and uniquely original each indi-vidual creature may appear to the average person, when all living things are grouped together according to their common anatomical features, the trend is unmistakable. Living things simply don't mix and match features; they fall into a pattern of distinct groups within groups, clearly demonstrating their an-cestral relationships.

Any discovery of a new species that didn't fit this pattern could poten-tially disprove the entire theory of evolution. For example, if birds were found with mammary glands or hair, if mammals were found with avian feathers or reptilian scales, if fish were found with mammalian cusped teeth, or if any other such things as these were found, the entire theory of common decent could come crashing down. So far, out of the approximately 1.5 million species that are currently known, there has not been even one found that does not fit the predicted pattern.[9] With less than 10% of all living things having been classified, there are still plenty of opportunities to falsify the theory of common descent by finding just one species that does not fit into the pattern of nested hierarchies.[10]

The only natural explanation for this is that all creatures within a group share a common ancestor, and that individual groups with similar features also share a common ancestor, all the way back to the first living cell. In other words, the design of each individual species appears to be entirely constrained by "family" relationships. I won't argue the fact that all living things clearly

---

[9] Douglas L. Theobald, *29+ Evidences for Macroevolution: The Scientific Case for Common De-scent*, The Talk.Origins Archive. Vers. 2.83.2004. 12 Jan, 2004, Part 1.2. http://www.talkorigins .org/faqs/comdesc/

[10] It has been suggested that a Panda's opposable "thumb" disproves *common descent* since this fea-ture, common to all primates, does not appear in any other bears. However, it can be demon-strated by anatomy that the Panda's thumb is actually an enlarged bone in the wrist called the *ra-dial sesamoid*, rather than an actual sixth digit. If a Panda's paw were to have four digits, a thumb, and a complete set of wrist bones as other primates have, then it would in fact be difficult to explain this in terms of evolution. But as it stands the anatomy of a Panda's paw is analogous to that of other bears, to which they are related.

demonstrate evidence of being designed by an intelligent agency. But in addition to showing evidence of *design*, each species also clearly shows evidence of *descent*, suggesting that a natural process is responsible for *assembling* these incredible designs of God.

It might seem strange at first to consider both *intelligent design* and *common descent*, but it would be similar to finding "Paley's Watch" in the wilderness, only rather than picking up a finely crafted Rolex, what we find is a hodgepodge of various watch parts salvaged from several older timepieces cobbled together as a monument to jerry-rigged engineering.[11] Sure, there had to be a designer, but he clearly must have been working within some obvious set of physical constraints. If we consider the evolutionary constraints to be the laws of nature working on pre-existing material, then we should be able to piece together the ancestral clues that each creature carries with it. In other words, the similarities and differences between groups of species should reveal their evolutionary histories; histories that would not exist if each species were created unique and descended only after its own kind.

Your Genes Don't Lie

Let's go back to the example of our five different generational sets of families. Suppose that all five families were from the Orient. Do you think you could still arrange them back into their family groups based solely on appearance? You might be able to, but it would be a lot more difficult. Similarities and differences in external appearances might still give you enough information, but wouldn't it be nice to have some other way to objectively and independently verify the family relationships? Well, you're in luck! The surest way to know whom has descended from whom is to perform DNA testing—something that modern technology easily affords us. Because DNA is always passed from parents to children, sequencing the genetic code of each individual should verify if the physical similarities are in fact due to common ancestry or are simply the result of coincidence.

In 1944 it was discovered that DNA was the universal genetic code for all living things. Since that time, the human genome, along with thousands of other species, has been sequenced and added to the GenBank.[12] So do you

---

[11] The Reverend William Paley (1743–1805) was an English theologian and philosopher who is best remembered for his watchmaker analogy. He articulately argued that just as a watch begs the question of a watchmaker, the existence of the creation is proof of a creator who acts outside of space and time.

[12] GenBank is an international, publicly accessible genes sequence databank containing information on hundreds of thousands of organisms. The amount of data is estimated to double every 10 months.

think that genetic testing was able to confirm the biological relationships that had been based solely on appearance? Would the theory of common descent pass the genetic test, or would DNA testing reveal that evolution was an elaborate hoax, perpetrated on a gullible public by unbelieving biologists who can't stand the idea of creation? I'll let you be the judge of that.

For those who may have forgotten their biology, a single segment of DNA consists of a long chain of molecules embedded with a sequence of chemical *bases* that creates a code for a specific protein. While there are over 99 functionally equivalent chemical bases found in nature, every living thing uses the exact same four bases.[13] With millions of possible combinations of bases, almost every known species could have had its own four. This would have been unmistakable proof of special creation. This means that God either created each species separate and distinct, but intentionally used the same four DNA bases when any combination would have worked, or it indicates that all living things share a common ancestor.

Any sequence of three DNA bases out of the four available possibilities is a code for a single *amino acid*. These amino acids are the building blocks of proteins, and proteins are the building blocks of cells. By stringing together long chains of amino acids according to the sequence of bases in the DNA code, proteins are formed with a characteristic shape that enables them to perform their intended functions. In addition to making up the cellular structures, proteins also regulate all functions within each cell.

There are over 320 naturally occurring amino acids. Since it only takes a sequence of three DNA bases to specify a single amino acid, the four DNA bases used by every living thing could specify any set of up to 64 (4^3) amino acids out of the 320 that are available. If each species were created separate and distinct, descending only after its own kind, you might expect to see different organisms using different sets of amino acids. Yet every known living thing uses practically the same set of 20 amino acids.[14] These other amino acids work just as good as the ones used, but no creature on earth uses them. Why is that? Again, God either intentionally created them to use the same 20 out of 320 possible amino acids, or every living thing has a common ancestor.

Since the genetic code could specify up to 64 out of the 320 naturally occurring amino acids but only 20 are used, it is a *redundant code*. What that means is that each of the 20 amino acids has more than one possible sequence

---

[13] Douglas L. Theobald, *29+ Evidences for Macroevolution: The Scientific Case for Common Descent*, The Talk.Origins Archive. Vers. 2.83.2004. 12 Jan, 2004, Part 1.1. http://www.talkorigins .org/faqs/comdesc/

[14] The slight deviations from the standard amino acid set typically involve a post-translation modification of a standard amino acid into a slightly different form. Ibid, Part 1.1.

of DNA bases that specifies it. The 64 different 3-base sequences that specify the 20 amino acids are known as the standard genetic code.[15] While there are over 10^{70} informationally equivalent genetic codes, all living things basically use the same one.

If each species were created separate and distinct, descending only after its own kind, you might expect to see some variation in the code. In fact, there are more than enough possibilities for every living thing to have its own unique genetic code. This would have actually been advantageous, eliminating things like viral infections. Moreover, if God had created each species with a unique genetic language, it would have been unmistakable evidence for special creation. Just think: the whole issue of evolution and common descent would have been settled 60 years ago—end of story! Instead, over 34 million new base sequences from various species of plants and animals are decoded every day. Each one has the potential to reveal something new, but no significant deviations from the standard genetic code have ever been discovered.[16]

In addition to the code itself, the basic metabolic pathways common to all livings cells such as *glycolysis*, the *citric acid cycle*, and *oxidative phosphorylation*, use the same steps in the same sequence with the same enzymes.[17] All of this despite the fact that there are numerous different possible metabolic pathways that achieve the same result. Now just think about that for a minute. Is there any reason why every plant, animal, fungus and bacteria must share the same genetic code, the same amino acids, and the same metabolic pathways? Perhaps this was all accomplished by special creation (ex nihilo), but the only *scientific* explanation is common descent.

The original question was how can DNA sequencing verify the pattern of nested hierarchies that was originally based on anatomical similarities? There just so happen to be ways to do this. Just about every living thing uses the same basic proteins for key metabolic functions. These proteins can be made

---

[15] A few of the 3-base sequences signify "start" and "stop" for each protein.

[16] There are a few changes to the standard molecular code found in mitochondria. As highly reduced genomes, they are more susceptible to modification. In hemichordates, one of the standard codons is not used at all in mitochondria. In echinoderms, close relatives of hemichordates, a mutation allows a different amino acid to use that "orphan" codon. Thus, this unique feature actually qualifies as a transitional form! See Jose Castresana, Gertraud Feldmaier-Fuchs, and Svante Pääbo, "Codon Reassignment and Amino Acid Composition in Hemichordate Mitochondria" *Proceedings of the National Academy of Sciences of the United States of America.* 1998 March 31; 95(7): 3703–3707.

[17] A metabolic pathway is a series of chemical reactions that takes place in a cell to produce something to be used or stored by the cell. Douglas L. Theobald, *29+ Evidences for Macroevolution: The Scientific Case for Common Descent*, The Talk.Origins Archive. Vers. 2.83.2004. 12 Jan, 2004, Part 1.1. http://www.talkorigins.org/faqs/comdesc/

up of hundreds or thousands of amino acids strung together in long chains. These chains fold over onto themselves based on what amino acids occupy certain key positions along the chains and the proteins take on a characteristic shape. The shape of a protein is what gives it its unique properties and allows it to perform its specified function. Because the universal genetic code is redundant, there are any number of ways to code for the same long amino acid sequence. Moreover, since some amino acids can be substituted without changing the overall shape or function of the protein, this adds even more possibilities for the DNA sequence that specifies the protein.

Mutations in a DNA sequence that produce no deleterious effects tend to accumulate over time since they are not weeded out by natural selection. These changes in the genetic code are passed to every new species from their ancestors, creating a distinct genetic pattern based on common ancestry. If the pattern of nested hierarchies (groups within groups) that was originally based solely on appearance is true, then the similarities and differences in the sequences of DNA that code for these common proteins should match the similarities and differences in anatomy. For instance, humans and chimpanzees should have more DNA sequences in common than humans and crocodiles, mammals and reptiles should have more DNA sequences in common than reptiles and insects, and so forth. And since the proteins analyzed are used in the basic metabolic pathways common to all living things, they should be able to give us results that are independent of the anatomical characteristics with which they have nothing do to.

This was a perfect test for evolution. Would the theory of common descent be finally proven false? Or would the picture shown by putting the genetic puzzle together match the pictures given by the fossil record and comparative anatomy? First of all, it's important to note that there are 10^{38} possible ways to arrange the 30 major biological groups of the standard "tree of life" into different patterns of nested hierarchies. If each species were designed and created separately, there should be no logical reason for the arrangement based on genetics to match the arrangement based on appearance. Moreover, there are so many different ways that a particular DNA sequence can code for the same protein that there is no reason to expect anatomically similar groups like humans and chimps to be any more genetically similar than humans and fish for the same basic proteins.

In other words, if the original parents of each living species were created ex nihilo by a miracle, then we should expect to see the Creator either using the same DNA sequence (to maximize efficiency), or using a completely different DNA sequence (to maximize diversity) when specifying functionally identical proteins. There should be no reason for certain species to use nearly identical sequences, while others use different sequences—unless the degree of similarity was a result of ancestral relationships (and 6,000 years is hardly enough time for a significant number of mutations to accumulate). However, when the genetic similarities and differences for the 30 major biological groups are compared, a striking pattern emerges.

In science, something is usually accepted without question if it can be measured out to at least three decimal places of accuracy. As it turns out, the measured statistical similarity of the genetic and morphological trees can be measured out to 38 decimal places![18] There are no other fundamental constants of nature that can be measured to that degree of accuracy. Once again, the only natural explanation of why genetics would tell us the same story as the fossil record and as comparative anatomy is that there must be some truth to the story.

Any supernatural explanation of this brings up a challenging question. If these genetic similarities and differences are based solely on design, and each species was created separately and only descends after its own kind, why would the Designer give us such an unmistakable appearance of common descent? Again, do Christians need to add the "appearance of evolution" argument to their "appearance of age" argument, just to maintain a literal scientific reading of Genesis?[19] If God created each species ex nihilo, what would be the point of purposely misleading us by designing creatures that appear to be genetically related by groups in a pattern of nested hierarchies that perfectly matches the pattern derived from comparative anatomy?

Interpreting Evolutionary History

All of this evidence is simply the fulfillment of a very important evolutionary prophesy: if all creatures are indeed related to one another by common ancestors in an unbroken tree of life all the way back to the first living cell, then the anatomy and genetics of all species should reveal their unique evolutionary histories—which is exactly what we find. Moreover, each of these lines of evidence (genetic, anatomical, geological, and geographical) should reveal a single consistent history for all of life. A history that, once identified, can either be accepted as authentic (evolution is a useful framework for understanding the facts of natural history) or dismissed as apparent (evolution is not a useful framework for understanding the facts of natural history). Christians should think carefully about how we respond to this.

If we apply the interpretive framework of special creation and assume that evolution is impossible, then these data have to be forced using unnatu-

---

[18] Ibid, Part 1.3.

[19] Terry M. Gray, "Biochemistry and Evolution", edited by Keith B. Miller, *Perspectives on an Evolving Creation* (Grand Rapids, MI; Eerdmans, 2003), pg. 263.

ral assumptions and awkward explanations. In that case, God is apparently using fossils, anatomy and genetics to deceive us. The conclusion is inescapable. But for what purpose would God try to trick us? Is this a test of our
faith? Would God ever require us to deny a pattern so clearly observed in nature? If so, then perhaps we should also deny the observations of modern astronomy and revert back to a belief in biblical geocentricity. Was that also a
test that we all failed? Perhaps the moon is really greater than the stars as
Moses clearly states? Have we been duped by the astronomers as well?

On the other hand, if we use an evolutionary framework to interpret these
data, then these anatomical and genetic relationships are completely natural.
They are exactly what we should expect to find if the theory of evolution has
any descriptive merit. Have we stumbled upon the method by which God creates through time? In this scenario, creation via evolution brings glory to God
while creation by divine fiat makes Him out to be a great deceiver. Now I ask
again: what advantage is there to using a supernatural framework to interpret
the data of natural history? Certainly Christians are free to take that approach,
but I just don't see how it profits us.

What about Intelligent Design?

To conclude that this evolutionary history is authentic doesn't remove
God as the grand architect of the cosmos any more than Newton's laws of motion remove Him as the sovereign governor of the material universe. But the
obvious evidence for "design" in the cosmos, even when attributed to the
handiwork of a supernatural intelligence, still requires some kind of material
mechanism to get from the ethereal realm of heaven down to the physical
world. Let us not confuse the transcendent concept of *design* with the material mechanism of *assembly*. Just as simple observation might lead one to
draw the design conclusion, the material mechanism of creation (assembly via
common descent) should also leave a trail of empirical evidence revealing its
physical history. And by all accounts, the testimony of nature suggests that if
intelligent design is true, it must have been accomplished within the constraints of a natural process—a process such as descent with modification
through gradual evolution.

Consider just a few more examples of this evolutionary history from the
biologists who study these things for a living:

> The tooth buds developed in the embryonic stage by birds and
> anteaters—buds that are later aborted and never erupt—are remnants
> of their toothed ancestors. The tiny vestigial wings hidden under the
> feathers of the flightless kiwi attest to its ancestors' ability to fly.

Some cave-dwelling animals have rudimentary eyes that cannot see, degenerate remnants of their ancestors' sighted ones.[20]

In addition to these, most pythons have a pelvis leftover from their legged ancestors. Dandelions reproduce without fertilization, but still produce pollen as a leftover feature from their flowering ancestors. Some flightless beetles have perfectly formed wings trapped beneath a hard exterior shell. Some whales are found with vestigial hind legs buried under layers of blubber resulting from previously inactivated genes that are accidentally "switched on" by a simple mutation.

Why would a whale have genes for hind legs that closely resemble those of a cow unless both of these creatures share a common ancestor? Any attempt to explain these structures in terms of special creation raises many awkward questions. Obviously God could have designed whales to have suppressed genes for bovine hind legs, but what would be the point of such a feature? Again, is God trying to intentionally confuse us? However, if God's divine plan to create seagoing mammals is achieved through an evolutionary process, then these anatomical vestiges make perfect sense.

The human body also shows many signs of suboptimal "design" that only make sense in terms of our evolutionary history. Human embryos have a yolk sac in the first stages that serves no apparent function and doesn't even contain any yolk. It is simply a curious relic of our reptilian ancestry—a heritage that is also confirmed by the fossil record, comparative anatomy, and molecular genetics. The human backbone, rather than being optimally configured for walking upright, clearly shows signs of being recently adapted from a pre-existing structure configured for walking on all fours. Why do you think we get so many back problems as we age?

Why do humans require vitamin C in their diets? Most other animals, except for the primates and guinea pigs, are fully capable of synthesizing vitamin C on their own. Curiously, the human DNA still contains all of the genes necessary to perform this function. However, the gene used in the last step of the process was "switched off" by a deleterious mutation of a single nucleotide nearly 40 million years ago. Because the primate diet already contained sufficient amounts of vitamin C, natural selection allowed this mutation to be passed on. As a result, all primates today still carry this same mutation to the same gene.

The odds of the same base deletion, out of a sequence spanning hundreds

---

[20] Jerry A. Coyne, "Intelligent Design: The Faith That Dare Not Speak Its Name" Edited by: John Brockman, *Intelligent Thought: Science Versus the Intelligent Design Movement* (New York, NY; Vintage Books, 2006), pp. 7-8.

of base pairs, occurring separately in all primate species is astronomical. The only reason why all primates would have this exact same mutation is if all primates shared a common ancestor. And the only reason all primates would still have nonfunctional copies of the genes required to synthesize vitamin C is if all primates shared a common ancestor with other mammals, whose genes were unaffected by this mutation.[21] Does any of this make sense in terms of special creation? Absolutely not! But in terms of creation by common descent, these kinds of things are to be expected.

There are many more observed instances of jerry-rigged adaptations in the human body. As a final example, consider the recurrent laryngeal nerve.

> ...a nerve that runs that runs from the brain to the larynx, helping us speak and swallow. In mammals, this nerve doesn't take a direct route but descends into the chest, loops around the aorta near the heart, and then runs back up to the larynx. It is several times longer than it needs to be; in the giraffe, the nerve has to traverse the neck twice and so is fifteen feet long—fourteen feet longer than necessary! The added length makes the nerve more susceptible to injury, and its tortuous path makes sense only in light of evolution. We inherited our developmental pathway from that of ancestral fish, in which the precursor of the recurrent laryngeal nerve attached to the sixth of the gill arches—embryonic bars of muscle, nerves, and blood vessels that developed into gills. During the evolution of land animals, some of the ancestral vessels disappeared, while others were rearranged into a new circulatory system. The blood vessel in the sixth gill arch moved backward into the chest, becoming the aorta. As it did so, the nerve that looped around it was constrained to move backward in tandem. Natural selection could not create the most efficient configuration because that would have required breaking the nerve and leaving the larynx without innervation.[22]

What are we to make of this? Most serious biologists, both Christian and non-Christian draw the same conclusion:

> These signs of history are the telltale marks of evolution, and all organisms have them. Because evolution can work only on the organism, structures, and genes that already exist, it seldom finds the per-

---

[21] The guinea pigs, which are not descended form primates, are also incapable of synthesizing vitamin C. As expected, this was caused by a *different* series of mutations, albeit to the same gene.

[22] Ibid, pg. 8.

fect solution for any problem. Instead, evolution tinkers, improvises, and cobbles together new organs out of old parts.[23]

With miracles, none of this "cobbling" would be necessary. Rather than having to reengineer structures and systems already present, an intelligent designer working through special creation would be free to start each new organism from scratch. If this is what actually happened, these "senseless signs" of evolutionary history should not exist. But they clearly do exist. What do we do about them?

Christians should not fear these kinds of discoveries. If we believe that God is the supreme governor of the cosmos, we already know that He is the conscious mind behind all of creation. But, all this evidence for common descent shows us that Intelligent Design theories make more sense if the actual mechanism of creation is material. Having each species appear "out of thin air" with a "built-in" evolutionary history that never actually happened only makes God a *deceptive designer*. This is ultimately no different than the "appearance-of-age" argument of Young-Earth Creationism.

I understand that many Christians will be uncomfortable with this. But what is the alternative? If we deny any possibility of evolution, then how should we interpret the evidence for common descent? We can't just ignore it by assuming we are not privy to God's secret purposes. There is nothing secret about this evidence! Every living thing carries its evolutionary history with it. We must therefore conclude that it is all a lie unnecessarily told by the Creator and Sustainer of the cosmos to either mislead or test us. By having God attempt to cover His creative tracks with an abundance of carefully crafted signs of evolutionary events that never took place, special creation takes a powerful ontological argument like *Intelligent Design* and turns it into *Intelligent Deception*. God has designed the universe with a specific intent to deceive us.

Is there another way to interpret the evidence for common descent that doesn't make God a liar, and doesn't remove Him as the architect of life? Why yes, there certainly is. It's simply called evolution, or as Darwin called it, descent with modification. You can even put a "theistic" qualifier in front if it makes you feel more spiritual. The evidence for common descent shows us that Darwin's theory of evolution, with its powerful combination of necessity (natural law) and contingency (chance) acting over billions of years, is just the kind of material mechanism of assembly that a sovereign creator might use to express His creative will on Earth.

---

[23] Kenneth R. Miller, *Finding Darwin's God: A Scientist's Search for Common Ground Between God and Science* (New York, NY; Harper Collins, 1999), pg. 101.

Will conservative Christians ever be able to embrace this? I believe so. It took hundreds of years for Christians to replace the Hebrew cosmos with the Greek version. A thousand years later, we struggled with replacing the geocentric model with the Copernican solar system. There is no turning back now. We've seen the machinery beyond the firmament. Eventually we will have to acknowledge it. But for now, there are some very big theological obstacles that must still be removed.

For instance, any time scientists refer to the laws of nature as being blind and purposeless Christians get nervous. Evolution does require a certain degree of "randomness" to provide the genetic variation necessary for natural selection to act upon. And most Christians get physically ill when entertaining any idea that all of life is contingent on undirected cause and effect. But it is a mistake to automatically equate this view of nature with atheism or materialism. Why should Christians see evolution through the lens of a godless worldview? If we know that God is the sovereign Creator and Sustainer of the universe, that He guides and directs the material cosmos according to His perfect patterns of providence, and that even seemingly contingent events governed by the laws of probability are under his divine command, how can any apparent randomness or chance be outside of God's control?

Dr. Francis Collins, head of the Human Genome Project, is an evangelical Christian and world-renowned scientist. He recently wrote the following:

> If God is outside of nature, then He is outside of space and time. In that context, God could in the moment of creation of the universe also know every detail of the future. That could include the formation of the starts, planets, and galaxies, all of the chemistry, physics, geology, and biology that led to the formation of life on earth, and the evolution of humans, right to the moment of your reading this book—and beyond. In that context, evolution could appear to us to be driven by chance, but from God's perspective the outcome would be entirely specified. Thus, God could be completely and intimately involved in the creation of all species while from our perspective, limited as it is by the tyranny of linear time, this would appear a random and undirected process.[24]

With a biblical theology of creation, many of the traditional evolutionary obstacles, such as "chance" can easily be overcome.

---

[24] Francis S. Collins, *The Language of God: A Scientist Presents Evidence for Belief* (New York, NY; Free Press, 2006), pg. 205.

The Real Challenge of Evolution

At this point, many readers are probably reaching deep into the recesses of their minds trying to recall every single reason why evolution is impossible. The objections are unending. What about the odds of complex systems arising spontaneously? How did all of the information get translated into the language of DNA? What about all of those creatures that "defy" a step-by-step evolutionary development? These are all good questions, and I'm sure biologists are working hard to solve them. There is so much wonder and mystery when it comes to creation that we may never have satisfying answers to these questions. But for Christians, the real challenges presented by evolution are not *scientific*. Our challenges are clearly *theological*.

Forget about science for a minute. I'll be the first to admit that the proposed evolutionary mechanisms can stretch the imagination. In fact, if the evolutionists didn't have such good poker faces, they would probably admit that these theories stretch their minds as well. Despite the converging lines of geological, anatomical, and genetic evidence that clearly indicate something along the lines of evolution *did* in fact happen, scientists still don't have many clues as to exactly *how* it could have happened. There are some impressive theories floating around out there, but let's face it: the proposed mechanisms of evolution can't be sufficiently demonstrated in a laboratory and will probably never win over the general public. Even movies like *Jurassic Park* have to speed up the evolutionary clock just so something interesting happens before the film ends.

We Christians could spend all of our time trying to prove how things like the bacterial flagellum or the human eye could not have possibly emerged by a natural mechanism. We could produce detailed calculations showing how the probabilities of evolutionary claims are too small to comprehend. We could break down the complex biochemical micro-machines inside of every living cell and demonstrate how such "irreducibly complex" things couldn't possibly exist apart from an intelligent Designer. We could do all of these things until we are blue in the face, and many of us have. But in reality, these issues are of absolutely no consequence to our faith. Who are we kidding? And neither can these things convince unbelievers to put their trust in Christ. These arguments against possible evolutionary mechanisms only distract us from the real challenges of evolution, which are clearly theological.

Consider this: if God is indeed governing the cosmos, and everything happens according to His divine plan (even those "random" events that we attribute to probability), then why should we care about the odds? In a world that is completely under God's command, what use is it for Christians to cal-

culate probabilities? The odds of the Red Sea parting just before the Israelites were about to cross it probably aren't any lower than the odds of the
bacterial flagellum evolving "naturally." The odds of a virgin becoming
pregnant aren't too great either. The odds of the Holy Scriptures being preserved without error over thousands of years after being copied and re-
copied by hand and candlelight are probably close to zero. In fact, the odds
are heavily stacked against every single thing that Christians hold sacred.
Do we really want to play the odds with the evolutionists? If anything, we
theists have an epistemological basis for believing the improbable, whereas
a materialist does not. Why would we throw this away and start arguing
against the odds?

Whether or not scientists ever figure out exactly how evolution happened,
there is more than enough evidence to show that it probably did. To use our
analogy from Chapter Seven, we don't know how the thief got into our house,
but all of our stuff is gone! Now what are we going to do about it? Should we
continue arguing over whose responsibility it was to lock the front door? Or
should we just deal with the situation as it presents itself? It's time for Christians to stop trying to argue the case that evolution is impossible. What's the
point? If evolution is indeed the way God chooses to create, and there is overwhelming evidence that He could have used this very process, then how could
it be impossible? With God all things are possible.

How then should we understand these data in a biblically and theologically consistent matter? Despite the physical evidence, some Christians will
still oppose evolution using Scripture. For example, Moses clearly tells us in
Genesis that each living thing "descends only after its kind." Could he have
been wrong? Of course not! While Moses makes the moon a "greater light"
than the stars, nobody today would try to turn this into an astronomical statement of fact. So why would we assume that when Moses describes creatures
descending after their kind, he is making a biological statement of fact? In the
words of John Calvin commenting on Genesis 1:6, "nothing is here treated of
but the visible form of the world. He who would learn astronomy (or biology)
and other recondite arts, let him go elsewhere."

What about the historicity of Adam and Eve? The many extinct hominid
species that appeared between the early apes and modern humans need to be
accounted for. A scientific interpretation of the fossil record assumes that these
creatures represent the evolutionary transitions from early ape to modern human. But the Bible clearly teaches that Adam and Eve were the first humans
created in the image of God from the "dust of the earth." To make matters even
more challenging, Paul links the fateful actions of Adam to the redeeming
work of Christ when he writes, "For if, by the trespass of the one man, death
reigned through that one man, how much more will those who receive God's

abundant provision of grace and of the gift of righteousness reign in life through the one man, Jesus Christ."[25] Is this a problem? You bet it is—and there are no easy answers to questions about Adam and Eve's historicity.

Like any other area of observational and historical science, there is no universal consensus on the forensic details of human origins. Recent advances in molecular genetics have enabled *paleoanthropologists*[26] to sequence DNA from the fossilized remains of extinct hominid species and compare it to modern humans. As a result, a strong, but not yet universal consensus on the natural history of man is beginning to emerge.[27] Most anthropologists believe that the first humanlike creatures appeared on the earth some 4.5 million years ago in northern Africa. While still more ape-like than human-like, these creatures were clearly built to walk upright, marking the separation of bipedal primates from the ancestors of modern apes.

Just as one would expect, assuming that the theory of evolution is true, these first hominid fossils are always found in the same geographic region as their ape-like ancestors. Of course, this fact could be explained in terms of creation by miracles, but the point is that it doesn't have to be. If these earliest hominids were found on any other continent, it might be a strong argument against evolution. But instead, the earliest humans are found right where one would naturally expect—on top of their nearest more ape-like ancestors.

Somewhere around two million years ago, these early African hominids began to use primitive tools made from chipped stone, indicating an increasing intelligence corresponding to their increasing brain size. The Neanderthal species were more similar to modern humans and are believed to have migrated from Africa to Europe about 250,000 years ago. Modern humans first appeared in North Africa about 200,000 years ago with slightly more advanced tools, but there is no indication of a distinctly *human* culture until about 40,000 years ago. Before that time, all indications are that the hominid creatures in North Africa lived and died like animals, with nothing more civilized than primitive tools made of chipped stone. But after that time, these people began to bury their dead, wear jewelry, make more elaborate and functional tools, and carve sculptures from teeth and bone.[28]

Clearly this behavior indicates that something significant happened around 40,000 years ago. People "suddenly" began to live differently than an-

---

[25] Romans 5:17

[26] A *paleoanthropologist* is somebody who studies the evidence for human evolution in the fossil record.

[27] Steve Olson, "Neanderthal Man," *Smithsonian,* Vol. 37, Number 7, October 2006, pp. 76-82.

[28] James P. Hurd, "Hominids in the Garden?" edited by Keith B. Miller, *Perspectives on an Evolving Creation* (Grand Rapids, MI; Eerdmans, 2003), pp. 208-233.

imals. They started to produce artwork which indicates a creative capacity. They appeared to have religious traditions and live in families. Could this be evidence of the image of God? If so, what about those earlier hominids? Were they human in form only, without souls?

During this well-documented explosion of cultural activity, these primitive North African humans spread out across the globe and built communities wherever they settled. Studies in molecular genetics indicate that they displaced the less advanced Neanderthal people of Europe with little or no interbreeding. There is some debate among anthropologists, but many think that the Neanderthal never developed culturally or technologically, which possibly contributed to their ultimate extinction after the influx of African peoples to Europe, about 30,000 years ago.[29] It appears that they simply couldn't compete with the more culturally and technologically advanced humans from North Africa.[30]

So what are we to make of all this? Where do Adam and Eve fit into this difficult scenario? Were they created miraculously? Certainly anything unknown can be explained with miracles, but how do we account for those human-like animals that lived and died millions of years before the image of God began to reveal itself in human culture? Were they biological ancestors to Adam and Eve? Did they have souls? Were they also capable of sin and did they require salvation? Many have speculated about this. C. S. Lewis said the following:

> For long centuries, God perfected the animal form which was to become the vehicle of humanity and the image of Himself... The creature may have existed in this state for ages before it became man: it may even have been clever enough to make things which a modern archaeologist would accept as proof of its humanity. But it was only an animal because all its physical and psychical processes were directed to purely material and natural ends. Then, in the fullness of time, God caused to descend upon this organism, both on its psychology and physiology, a new kind of consciousness which could say "I" and "me," which could look upon itself as an object, which knew God, which could make judgments of truth, beauty and goodness... We do not know how many of these creatures God made, nor how long they continued in the Paradisal state. But sooner or later

---

[29] There is now some evidence that the last Neanderthals did show cultural change, possibly by imitating modern humans with whom they briefly coexisted.

[30] David Wilcox, "Finding Adam: The Genetics of Human Origins," edited by Keith B. Miller, *Perspectives on an Evolving Creation* (Grand Rapids, MI; Eerdmans, 2003), pp. 234-253. There are several references from this essay throughout this section of the chapter.

they fell... For all I can see, it might have concerned the literal eating of a fruit, but the question is of no consequence.[31]

Is Lewis right? Perhaps. When it comes to an evolutionary understanding of natural history, these are the complex issues that Christians need to deal with, not probability or information theory.

It's a waste of time and energy trying to show how complex structures like the human eye or the bacterial flagellum couldn't have possibly evolved by a natural mechanism. What is the point of haggling over the possible evolutionary mechanisms of such relatively minor features when the entire 4.5 million-year history of human evolution from bipedal apes to modern humans, and the 50 thousand-year history of human migration from Africa to South America can all be documented with archeological and genetic evidence? Moreover, if these complex structures really can prove that evolution never happened—then what? The positive evidence for common descent doesn't just conveniently go away. We are then left with evidence of a Creator that is nothing more than a half-witted deceiver who can't even maintain a consistent charade. Is that the theological price of "winning" the argument against possible evolutionary mechanisms?

As more and more human fossils are unearthed, more pieces of the puzzle of man's natural history are put together. The science of molecular genetics is enabling DNA to be successfully sequenced from more fossilized remains. This is giving paleoanthropologists even more ways to trace the history of man's natural origins. Eventually, there may be enough information to provide more insightful answers to these serious questions. Are Christians ready for that?

Despite these and other challenges, the big advantage to giving up on creation science is that these serious issues do not have to be hastily resolved with half-baked solutions that satisfy neither sound theology nor honest science. We are perfectly content to let them remain mysteries until more clues are revealed to us. Not every difficult question is going to have a satisfying answer this side of eternity. Can our faith handle that? Do we really need detailed explanations for all the mysteries of Christian spirituality? I sure hope not. Christians seem to be perfectly content to rest in the fact that God is *one*, but exists in *three* persons; that Jesus was fully *God* and yet fully *human*; and that while God is perfectly sovereign over every detail of creation, He is mysteriously neither the author of, nor is He responsible for the evil acts of men.

If we can tolerate these unsolved theological mysteries, then why is there such a pressing need to have scientific answers for the unsolved mysteries of

---

[31] C. S. Lewis, *The Problem of Pain* (New York, NY; Simon and Schuster, 1996), pp. 68-71.

creation? Consider this from a Christian biologist who clearly takes these kinds of questions seriously:

> I want to close by pointing out that it is in this sort of situation where the "rubber really meets the road" in questions of faith and science. The question is the matter of data. Good theology cannot call James "an epistle of straw" when the verses don't fit its theological model. Likewise, good science cannot neglect data that do not fit its theoretical model. It must include and explain everything, or it must leave room for mystery. So, in this case, should I cobble together an "integrative" solution to resolve the tension, or should I wait on the Lord to resolve the issue in his own time through more data: Clearly the Lord knows the answer to these questions. I don't. If I have the faith to walk in obedience, should I be willing to wait with unresolved questions: Or should I insist on an immediate answer? What do you think?[32]

Creation Science and Apologetics

One of the main reasons that Christians take such an interest in Creation Science is to defend their faith. The other side of the apologetical coin is to promote the Christian faith by convincing others that God is real, Jesus is His son, and the Bible is His Word. But is Creation Science an effective tool of evangelism? Do skeptics, atheists and agnostics reject the grace of God for lack of scientific evidence? I submit to you that they do not. They don't need cognitive arguments, they need saving grace.

The claims of Christ will not make any sense when filtered through the materialist worldview. Without first acknowledging the existence of a sovereign Lord, any evidence of creation miracles will be rationalized away. If they will not believe the miracle of Christ's death and resurrection, do we honestly expect them to believe that the sudden appearance of species in the fossil record is evidence of God despite the overwhelming evidence for common descent? Or do we expect them to admit that their inability to explain the natural origin of a complex biochemical feature is proof of the existence of an intelligent Designer?

Consider these words of wisdom from a Christian professor of biology:

Miracles are at best a crutch for those, like Thomas, of little faith. Miracles by themselves don't convince a skeptic to become a be-

---

[32] David Wilcox, "Finding Adam: The Genetics of Human Origins," edited by Keith B. Miller, *Perspectives on an Evolving Creation* (Grand Rapids, MI; Eerdmans, 2003), pg. 253.

liever, and their absence shouldn't dissuade a true believer. In the terminology of logic, miracles are neither necessary nor sufficient for belief in God.

Focusing on tiny miracles like the bacterial flagellum wind up diverting us from the miracles in scripture...It turns you away from the Bible to look at scientific data. If you don't believe the Bible's miracles, check out the bacterial flagellum. Instead, all of creation is the miracle, not bits and pieces here and there.[33]

The only thing capable of melting down the wall of spiritual denial that surrounds the hearts of unbelievers is the grace of God. What the world wants to see is Christians putting their spiritual "money" where their mouths are. They are not impressed when Christians try to impersonate scientists by empirically searching for signs of a Creator, but they would be taken aback if Christians were to simply act like Jesus by performing genuine acts of Christian mercy and compassion with no theological strings attached. Our labors for the Kingdom of God should cut right down to the chase, serving our fellow men and women in need by tangibly extending the grace of God to a fallen and hurting world.

As wonderful as the natural sciences are by enhancing our quality of life and giving us a window into creation, they don't hold a candle to the simple ordinary acts of kindness and compassion that the followers of Christ are commanded to perform. Extending a cup of cold water to a hot and thirsty atheist will have a greater impact than all of the lofty apologetical musings of Young-Earth Creationism, Old-Earth Creationism and the Intelligent Design movement put together. How many brilliant scientific minds have already been isolated from the gospel message by misguided and dogmatic approaches to theories of natural history? How many valuable Christian resources have been wasted by groups like the *Institute for Creation Research* or *Answers in Genesis* to fight an unnecessary battle that can't be won?

In the end, Christians are far better off treating naturalistic theories about origins the same way we treat the rest of the natural sciences; as tentative material explanations of observable physical phenomena—nothing more, and nothing less. As long as the Christian worldview is anchored in a biblical theology of creation that finds God providentially working through all of the normal patterns of material behavior to accomplish His divine purposes, no scientific theory can possibly dethrone God as the Sovereign Lord of heaven and earth.

---

[33] Joan Roughgarden, *Evolution and Christian Faith: Reflections of an Evolutionary Biologist* (Washington, DC; Island Press, 2006), pg. 98.

CPSIA information can be obtained at www.ICGtesting.com
Printed in the USA
LVOW10s1313301213

367442LV00013B/149/P